STRATHCLYDE UNIVERSITY LIBRARY

30125 00331991 9

Computers in Design, Construction and Operation of Automobiles

Editors:

T.K.S. Murthy
C.A. Brebbia

A Computational Mechanics Publication

Springer-Verlag Berlin Heidelberg New York
London Paris Tokyo

T.K.S. MURTHY
Consultant
Computational Mechanics Institute
52 Henstead Road
Southampton
SO1 2DD
U.K.

C.A. BREBBIA
Computational Mechanics Institute
52 Henstead Road
Southampton
SO1 2DD
U.K.

Sub-editors: C.M. Mellors and J. Knudsen

 British Library Cataloguing in Publication Data

    Computers in design, construction and operation of automobiles.
    1. Automobiles — Design and construction — Data processing
    I. Murthy, T.K.S. II. Brebbia, C.A.
    629.2'3'00285   TL240

    ISBN 0-905451-79-1

ISBN 0-905451-79-1 Computational Mechanics Publications Southampton
ISBN 0-931215-69-2 Computational Mechanics Publications Boston Los Angeles
ISBN 3-540-17503-2 Springer-Verlag Berlin Heidelberg New York London Paris Tokyo
ISBN 0-387-17503-2 Springer-Verlag New York Heidelberg Berlin London Paris Tokyo

This work is subject to copyright. All rights are reserved, whether the whole or part of the material is concerned, specifically those of translation, reprinting, re-use of illustrations, broadcasting, reproduction by photocopying machine or similar means, and storage in data banks. Under §54 of the German Copyright Law where copies are made for other than private use, a fee is payable to 'Verwertungsgesellschaft Wort', Munich.

© Computational Mechanics Publications 1987
   Springer-Verlag Berlin, Heidelberg
   Printed in Great Britain.

The use of registered names trademarks etc. in this publication does not imply, even in the absence of a specific statement, that such names are exempt from the relevant protective laws and regulations and therefore free for general use.

PREFACE

The advantages of computer-aided systems in the automotive industry are now widely recognised. Computers are being used from the initial conceptual design and feasibility study to the actual construction and ultimate operation of the cars on the road. In the field of car body design, for example, the use of computers is replacing the traditional, if more expensive, testing methods, such as in a wind tunnel, due to the extensive development of numerical computational methods.

This book contains some of the papers presented at the 1st International Conference on Computer Aided Design, Manufacture and Operation in the Automotive Industries (COMPAUTO 87) organised by the Computational Mechanics Institute of Southampton, England, and held in Geneva from 10-12 March 1987. It excludes presentations dealing specifically with structural design and crashworthiness, which are published simultaneously in the companion volume entitled "Structural Design and Crashworthiness of Automobiles".

The papers contained in this book relate to the use of computers in such diverse applications as solid modelling, computational fluid dynamics, computer simulation, engine dynamics and other topics in design. Papers on manufacture and operation deal with sheet metal forming, robots, performance testing, vehicle and road simulation, noise levels, CAD data communication and data processing as well as CAE systems.

The book contains descriptions of some valuable engineering tools which will be of interest to designers, manufacturers and operators of automobiles.

T.K.S. Murthy
C.A. Brebbia

CONTENTS

SECTION 1 AERODYNAMIC SIMULATION

Status and Requirements for Numerical Aerodynamic Simulation in Automotive Industry
B. Wagner, W. Schmidt — 3

The Complex Variable Boundary Element Method Applied to Automotive Streamlining
T.V. Hromadka II, C.C. Yen — 11

The Solution of the Reynolds Averaged Navier-Stokes Equations in General Curvilinear Coordinates and its Application to Vehicular Aerodynamics
R. Patel and S. Dvinsky — 29

SECTION 2 ENGINE DESIGN

Models for Predicting Engine Torque Response During Rapid Throttle Transients in Spark Ignition Engines
I.C. Finlay, D.J. Boam, J.J.G. Martins, A. Gilchrist and C.K. Lee — 47

An Optimisation of Characteristic Parameters of Triple Stage Non Linear Torsional Damper in Automotive Clutch
Shao Cheng and Lu Zhenhua — 73

SECTION 3 MANUFACTURE

A Numerical Tool for CAD of Sheet Metal Forming Process in Automotive Industry
B.İ. Kılkış — 89

Computational Verification of Numerical Control Programs for Sculptured Surface Parts
J.H. Oliver and E.D. Goodman — 105

A Vibratory Device for Locating Objects: Theory and Experimental Results
D.T. Pham and J. Menéndez — 121

SECTION 4 PERFORMANCE

Programmable Instrumentation Recorder and Support Computer for Racing Car Performance Testing
M.F. Bessant — 141

Route Simulation as a Product Development Tool 159
R.A. Buré

Calculation of Noise Levels in Vehicle Interiors 179
H. Goldstein

SECTION 5 CAD DATA COMMUNICATION/CAE SYSTEMS

E.D.I. – Easy Direct Information 193
J.P. Newton

CAD Data Communication 207
M.J. Newton, K.S. Hurst

Application of CAD to Autobody Design in a Commercial Multi-Client Environment 215
D.J. Clark

CAE: Fact or Fiction 233
A Historical Review within an Automotive Company
W.R. Buell

Experience with an IEEE 802.3 Local Area Network in a Multi-Vendor Environment 247
W. Schmatz

Applications of Voice Recognition Techniques in the Automotive Industry. (Austin Rover Group. England) 257
R.P. Anderson, K.D. Gill

SECTION 1 AERODYNAMIC SIMULATION

Status and Requirements for Numerical Aerodynamic Simulation in Automotive Industry
B. Wagner, W. Schmidt
Aerodynamics/Theoretical Aerodynamics, Dornier GmbH, D-7990 Friedrichshafen 1, Federal Republic of Germany

INTRODUCTION

Since a couple of years a continuously increasing demand for numerical methods for computation of aerodynamic flowfields around car configurations has been observed. While the main interest up to now was dedicated to the overall prediction of the mechanisms creating aerodynamic drag including the flow separation in different regions of the car surface, recently the attention of car designers is also drawn to other basic aerodynamic problems. Expecially it is understood that yaw effects are important with respect to lateral driving stability, that the aerodynamic interference of the car bottom with the ground - either fixed ground referring to the usual wind tunnel situation or moving ground representing the car on the road case - cannot perfectly be analysed in wind tunnels, and that the interaction between aerodynamic loads and structural deformations can become essential in some regions. Furthermore, the internal aerodynamics of the unsteady flow in piston engines including not only the in-cylinder flow but also the flows through attached tube systems is a challenge for numerical methods which is far from being satisfactorily solved in spite of some existing preliminary computer programs since none of them resolves the viscous effects sufficiently by the coarse meshes used.

COMPUTATIONAL METHODS AND RESULTS

The coupled use of panel methods for the external inviscid potential flow and of boundary layer methods for the viscous effects at the body surface is already a powerful tool practically used in car aerodynamics. The procedure has got widespread applications, especially for the forebody design and optimization. The corresponding program package developed at Dornier uses a first order panel method and an integral boundary layer code which is fully three-dimensional in order to provide realistic results with respect to the viscous effects on the body surface. It allows for iterative application of both calculation steps by use of the equivalent source concept for simulating the displacement effects of the boundary layer in the inviscid flow solution. Even the effects of a boundary layer on a wind tunnel ground plate can be introduced by this iterative treatment.

The key for achieving optimal solutions for the flow around car configurations with panel methods is a simple but satisfactory model of the large separated flow regions behind the car. Dornier introduced an easy and very successful procedure by using a first order method with pure source distributions on the surface panels but omitting a panelization of the blunt base area. Therefore a flow out of the base region develops automatically adjusting the flow to the last panels' direction while an additional panelization of the base surface would enforce a rear stagnation point. Fig. 1 shows considerable pressure differences between both solutions at the upper surface reaching up to the windscreen. Hence, only by the open base excellent agreement [1] can be expected with experiments over a large portion of the upper surface as shown in Fig. 2. Unsatisfactory agreement can often be observed beneath the car, even if the effect of the boundary layers is iteratively taken into account, since this region is blocked in real flow by viscous effects leaving no space for an inviscid core flow (Fig. 3)[2]. By inspecting the streamwise boundary layer development (Fig. 3) the danger of separation can easily be judged by the accumulation of boundary layer material and the corresponding changes in the boundary layer shape parameter. But it must be emphasized that the viscous effects can only be reliably analyzed if a fully 3D boundary layer method is used.

Recently, besides the challenging task of predicting causes and magnitude of the aerodynamic drag, the importance of other aerodynamic force and moment coefficients is recognized with respect to driving stability of all kind of cars including especially the performance of racing cars. Correspondingly, it will be mentioned that the above program package reveals even for asymmetric flow conditions at small yaw angles reasonable results over the main part of the body, excluding only the separated region at the very end of the body.

Numerical solutions of the Euler equations, which represent the exact equations for compressible inviscid flow have become very common since a couple of years. Such solutions can more realistically model the flow about car shapes since these equations allow for correct vorticity convection while the correct vorticity production at solid walls can be provided by coupling the inviscid Euler method with a boundary layer method. At Dornier a block structured mesh generation strategy has been developed in order to solve the time-dependent Euler equations by a finite volume method for the three-dimensional flow about arbitrary complex shapes. Due to the block structure the number of mesh points is principally not limited by the available main storage capacity of the actual computer. The calculations are very economic because the convergence to steady state is speeded up by some acceleration features including a multigrid strategy.

The Euler methods represent a preliminary step in solving the Navier-Stokes equations by the use of the same numerical scheme. They can deliver very valuable results which may include the afterbody flow, especially if coupled with a 3D boundary layer code. We solve the nonlinear time-dependent equations by a finite volume scheme using multi-stage time-stepping for timewise integration. We prefer a finite volume approach since it guarantees an optimal representation of physics based on the integral form of the conservation laws for mass, momentum and energy including correct treatment of discontinuities. Its application is convenient since it permits contour conformal meshes with simple Cartesian representation of coordinates and momentum components, it avoids numerical differentiation as far as possible, and it is very forgiving with respect to mesh inhomogeneities (see e. g. [3,4]). Fig. 4 shows the three-dimensionally calculated pressure distribution in the car mid plane plotted normally to the surface [5] and compared to experimental results. Considerable agreement between theory and experiment is shown over the upper surface except at highly curved regions where the mesh of about 200 000 grid points is still too coarse. In addition at the front portion of the car a block edge is located just at the highest curvature position deteriorating locally the quality of solution, which can be overcome by changing the block structure. Beneath the car the disagreement is obviously due to considerable viscous effects at low ground clearance as already discussed above with panel methods. Theory and experiment disagree at the base because real viscous effects are decisive for the base pressure. Even in this inviscid Euler analysis a wake of qualitatively realistic structure occurs which cannot be quantitatively correct but often can guide the understanding of the vortex formation observed experimentally. Quantitatively correct computational results, especially with respect to the determination of the pressure distribution in the separated region, demand Navier-Stokes analysis with sufficiently fine meshes.

A Navier-Stokes method has also been derived by extension of the Euler method using essentially the same numerical scheme [6]. A two-dimensional version has been extensively used in a block structured form for 2D car shapes [7]. An algebraic two layer eddy viscosity turbulence model is implemented, namely the model of Baldwin-Lomax [8]. Thorough investigations have been made aiming at the comparison of viscous and inviscid (laminar and turbulent) flows and revealing the effect of numerical errors (numerical viscosity) compared with the real viscous effects. Also interesting results can be achieved which simulate the differences between wind tunnel experiments, where the ground plate is fixed with respect to the car, and the real road situation, where the ground is moving with respect to the car. Fig. 5 shows a comparison of calculated 2D-streamlines for both situations [8] revealing some essential differences which can affect the whole flow field.

From the result of this 2D pilot code it can be concluded that 3D solutions for reliable and sufficiently detailed results need considerable efforts with respect to mesh resolution and numerical quality of the scheme. Having these decisive factors in mind and comparing existing 3D Navier-Stokes solutions it cannot be judged up to now which role the turbulence model finally will play when all other problems have been satisfactorily solved.

Since the Navier-Stokes code developed at Dornier basically solves the time-dependent equations in order to perform the iterative solution procedure by time-stepping, the simulation of time-dependent processes is easily possible. Fig. 6 presents results of an axisymmetric Navier-Stokes calculation of the compression stroke in a Diesel engine clyinder having a considerable swirl prescribed initially [9]. In the present solution for the first time to our knowledge a trial was made to resolve the actual boundary layers developing at the cylinder walls while previous solutions used the law of the wall for the corresponding regions. Some interesting flow details were revealed during earlier crank angles but the final results are very similar to solutions by other well known methods. For achieving correct viscous solutions, especially e. g. the details of vortex formations around an inlet valve during a suction stroke, much finer mesh resolutions have to be used and much more efforts have to be spent for getting quantitatively reliable results.

A recent more detailed discussion of the above methods, their practical use and economy as well as their limitations can be found in Ref. [10].

REFERENCES

1. Seibert, W., Panelization Variations at a Car Afterbody, Unpublished Work at Dornier, 1984

2. Seibert, W., Aerodynamische Nachrechnung der Daimler-Benz-Fahrzeugkonfiguration C-111, Dornier Report 82 BF/16 B, 1982

3. Jameson, A., Schmidt, W., Some Recent Developments in Numerical Methods for Transonic Flows, Comp. Meth. in Applied Mech. and Eng. 51, 1985, 467-493

4. Wagner, B., Leicher, S., Schmidt, W., Applications of a Multigrid Finite Volume Method with Runge-Kutta Time Integration for Solving the Euler and Navier-Stokes Equations, 2nd European Conference on Multigrid Methods, Cologne, 1985

5. Fritz, W., Leicher, S., Seibert, W., Grid Generation, Euler Solution, and Iterative Boundary Layer Calculation for a Passenger Car, Unpublished Work at Dornier, 1985

6. Haase, W., Wagner, B., Jameson, A., Development of a Navier-Stokes Method Based on a Finite Volume Technique for the Unsteady Euler Equations, Proc. 5th GAMM Conf. on Numerical Methods in Fluid Mechanics, Rome, 1983

7. Fritz, W., Two Dimensional Euler and Navier-Stokes Solutions of Flow Over the Mid Section of a Car, 2nd IAVD Congress on Vehicle Design and Components, Geneva, Switzerland, 1985

8. Baldwin, B. S., Lomax, H., Thin Layer Approximation and Algebraic Turbulence Model for Separated Flows, AIAA 78-257, 1978

9. Haase, W., Misegades, K., Wagner, B., Calculation of Axisymmetric in Cylinder-Flow in a Diesel Engine during a Compression Stroke with Initial Swirl, unpublished Work at Dornier, 1985

10. Wagner, B., Schmidt, W., Computation of Automobile Aerodynamics by Use of Numerical Methods Developed in Aeronautical Industry, Proceedings of 3rd Autotechnologies Monte Carlo Conference, 1987

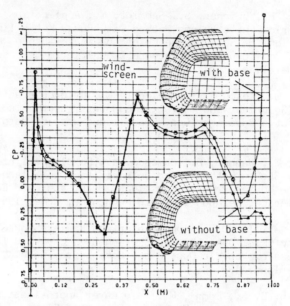

Fig. 1: Pressure Coefficient from Panel Calculation - Mid Plane, Upper Surface (2,5 % Ground Clearance)

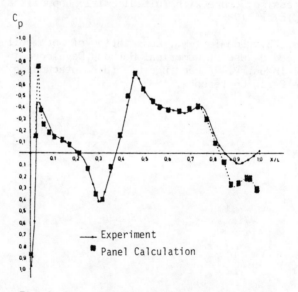

Fig. 2: Comparison between Pressure Coefficients Calculated by Panel Method and Experimental Results - 2,5 % Ground Clearance

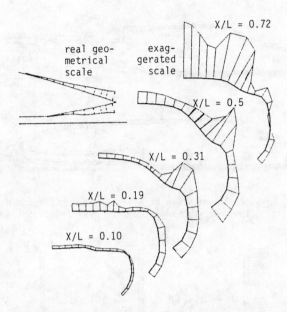

Fig. 3: Three-Dimensional Displacement Thickness
3D Panel-Boundary Layer approach

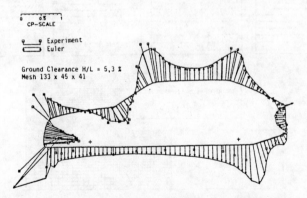

Fig. 4: Comparison between Pressure Coefficients Calculated by Euler Method and Experimental Results - Symmetry Plane -

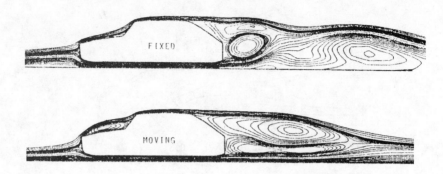

Fig. 5: Navier-Stokes Simulation of Fixed (Wind Tunnel) and Moving (Road) Ground Situation - Streamlines from Navier-Stokes Computations

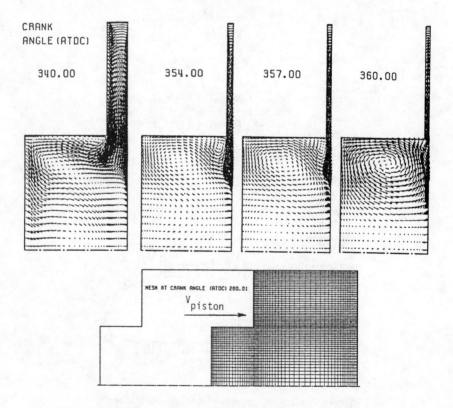

Fig. 6: Flow Field after Compression Stroke in an Axisymmetric Diesel Engine Cylinder with Initial Swirl (60 m/s Maximum Piston Speed)

The Complex Variable Boundary Element Method Applied to Automotive Streamlining

T.V. Hromadka II, C.C. Yen
Williamson & Schmid, 17782 Sky Park Boulevard, Irvine, CA 92714, U.S.A.

INTRODUCTION

The Complex Variable Boundary Element Method or CVBEM has been shown to be a useful tool for the numerical analysis of Laplace or Poisson equation boundary value problems (Hromadka, 1984a). The numerical procedure is to discretize the boundary Γ by nodal points into boundary elements, and then specify a continuous global trial function $G(\zeta)$ on Γ as a function of the nodal values. Using the Cauchy integral, the resulting integral equation is

$$\hat{\omega}(z_0) = \frac{1}{2\pi i} \int_\Gamma \frac{G(\zeta) d\zeta}{\zeta - z_0} \qquad (1)$$

where $\hat{\omega}(z_0)$ is the CVBEM approximation for $z_0 \in \Omega$; and Ω is a two-dimensional simply connected domain enclosed by the simple closed contour Γ.

Because $G(\zeta)$ is continuous on Γ, then $\hat{\omega}(z)$ is analytic over Ω and can be rewritten as the sum of two harmonic functions

$$\hat{\omega}(z) = \hat{\phi}(z) + i\hat{\psi}(z) \qquad (2)$$

Thus both $\hat{\phi}(z)$ and $\hat{\psi}(z)$ exactly satisfy the Laplace equation over Ω.

Approximation error occurs due to $\hat{\omega}(z)$ not satisfying the boundary conditions on Γ exactly. However, an approximative boundary $\hat{\Gamma}$ can be developed (by trial and error) which represents the location of points where $\hat{\omega}(z)$ does equal the specified boundary conditions such as level curves (see Fig. 1). Consequently, the CVBEM approximation error can be interpreted as a transformation of $\Gamma \to \hat{\Gamma}$ where the ultimate objective is to have $\hat{\Gamma}$ coincident with Γ. Because all the error of approximation is due to the incorrect boundary element trial functions, accuracy is increased by the addition of boundary nodal points where approximation error is large (i.e., adaptive integration).

In this paper, a computer interactive technique is reported which graphically displays Γ and $\hat{\Gamma}$ so that the numerical analyst can readily specify additional nodal points on the CRT screen. In this fashion, the user interacts with the CVBEM to locate the necessary nodal point additions until $\hat{\Gamma}$ and Γ are within an acceptable level of tolerance. For example, the tolerance may be the allowable construction limits specified for a shaft (torsion problem) for use in aircraft design.

As $\hat{\Gamma}$ approaches Γ geometrically, the analyst is assured by the Maximum Modulus Theorem that the maximum approximation error occurs on Γ and that the governing partial differential equation (Laplace) is solved exactly. Consequently, the final product is the exact solution for a problem geometry which is within the construction tolerance of the design.

THEORETICAL BACKGROUND OF THE CVBEM

A complete presentation of the CVBEM development, case studies, mathematical proofs of convergence and existence, and several FORTRAN computer programs are given in Hromadka (1984a). In order to develop the geometric interpretation of modeling error associated with the approximative boundary concept, a brief development of the CVBEM numerical technique is presented in the following.

Let Ω be a simply connected two-dimensional domain (i.e. no holes within Ω) enclosed by a simple closed contour Γ (e.g. Mathews, 1982). Let $\phi(x,y)$ be a two-dimensional harmonic function over $\Omega \cup \Gamma$; that is,

$$\frac{\partial^2 \phi(x,y)}{\partial x^2} + \frac{\partial^2 \phi(x,y)}{\partial y^2} = 0 \, , \quad (x,y) \in \Omega \cup \Gamma \qquad (3)$$

Then there exists a simply connected domain Ω^* such that $\Omega \cup \Gamma$ is a proper subset of Ω^* and $\phi(x,y)$ is harmonic over Ω^*.

There exists a harmonic function $\psi(x,y)$ conjugate to $\phi(x,y)$ which also satisfies the Laplace equation of (3) over Ω^* and additionally satisfies the Cauchy-Rieman conditions of

$$\frac{\partial \phi(x,y)}{\partial x} = \frac{\partial \psi(x,y)}{\partial y} \, , \quad \frac{\partial \phi(x,y)}{\partial y} = - \frac{\partial \psi(x,y)}{\partial x} \qquad (4)$$

Let $z = x + iy$ be a complex variable over Ω^*. Then both $\phi(x,y)$ and $\psi(x,y)$ can be written in terms of $\phi(z)$ and $\psi(z)$ such that an analytic function $\omega(z)$ is defined over Ω^* by

$$\omega(z) = \phi(z) + i\psi(z) \tag{5}$$

where to simply notation, (5) can be rewritten as $\omega = \phi + i\psi$, $z \in \Omega^*$.

Equation (5) represents a relationship between two conjugate harmonic functions generally called the potential (ϕ) and stream functions (ψ). A list of typical potential and stream functions which occur in engineering and physics is given in Table I (Mathews).

TABLE I. POTENTIAL AND STREAM FUNCTIONS

Physical Phenomemon	$\phi(x,y)$ = constant	$\psi(x,y)$ = constant
Heat flow	Isothermals	Heat flow lines
Electrostatics	Equipotentials	Flux lines
Fluid Flow	Equipotentials	Stream lines
Gravitational field	Potentials	Lines of force
Magnetism	Potentials	Lines of force
Diffusion	Concentration	Lines of force
Elasticity	Strain	Stress lines
Current flow	Potential	Lines of force

The Cauchy integral theorem equates values of $\omega(z_0)$ for $z_0 \in \Omega$ to a line integral of $\omega(\zeta)$ for $\zeta \in \Gamma$ by

$$\omega(z_0) = \frac{1}{2\pi i} \int_\Gamma \frac{\omega(\zeta) d\zeta}{\zeta - z} \tag{6}$$

To illustrate the development of a CVBEM approximation function, $\hat{\omega}(z)$, consider $\omega(z)$ to be defined over Ω^* with $\Omega \cup \Gamma$ interior of Ω^*. Subdivide Γ into m boundary elements Γ_j such as shown in Fig. 2. Nodal points are specified at each element endpoint (here, a linear polynomial CVBEM approximation is being developed). At each node, determine nodal values of $\omega(z)$ by

$$\omega(z_j) \equiv \omega_j = \phi(z_j) + i\psi(z_j) \equiv \phi_j + i\psi_j\,;\quad j = 1,2,\cdots,m \tag{7}$$

14 COMPLEX VARIABLE B E METHOD

Then a global trial function of $\omega(z)$ is determined for $z \in \Gamma$ by

$$G(z) = \sum_{j=1}^{m} \delta_j [\omega_j N_j(z) + \omega_{j+1} N_{j+1}(z)] \qquad (8)$$

Where the $N_j(z)$ are linear basis functions (see Fig. 3); and $\delta_j = 1$ for $z \in \Gamma_j$, and $\delta_j = 0$ for $z \notin \Gamma_j$. Substituting $G(z)$ in place of $\omega(\zeta)$ in (6) determines a CVBEM approximation $\hat{\omega}(z)$ of $\omega(z)$

$$\hat{\omega}(z) = \frac{1}{2\pi i} \int_\Gamma \frac{G(\zeta) d\zeta}{\zeta - z} \qquad (9)$$

Letting $||\Gamma_m|| = \max |z_{j+1} - z_j|$, $j = 1, 2, \cdots, m$, then it is seen (without proof) that

$$\lim_{||\Gamma_m|| \to 0} G(\zeta) = \omega(\zeta), \quad \zeta \in \Gamma \qquad (10)$$

and therefore

$$\lim_{||\Gamma_m|| \to 0} (\omega(z) - \hat{\omega}(z)) = \lim_{||\Gamma_m|| \to 0} \frac{1}{2\pi i} \int_\Gamma \frac{(\omega(\zeta) - G(\zeta)) d\zeta}{\zeta - z} = 0 \qquad (11)$$

Thus the error of approximation, $e(z)$, is defined by

$$e(z) = \frac{1}{2\pi i} \int_\Gamma \frac{(\omega(\zeta) - G(\zeta)) d\zeta}{\zeta - z} \qquad (12)$$

Because $G(\zeta)$ is continuous on Γ then $\hat{\omega}(z)$ is analytic over Ω which implies both $\hat{\phi}(z)$ and $\hat{\psi}(z)$, where $\hat{\omega}(z) = \hat{\phi}(z) + i\hat{\psi}(z)$, are potential functions over Ω.

In practice, $\phi(z)$ is known on Γ_ϕ and $\psi(z)$ is known on a separate contour on Γ_ψ where $\Gamma = \Gamma_\phi \cup \Gamma_\psi$. Thus $\hat{\omega}(z)$ is not completely defined without estimates for the unknown nodal values. To obtain such estimates, the real (or imaginary) parts of $\hat{\omega}(z)$ are collocated to the m known nodal values, resulting in m equations for the m unknown nodal values. Using these m nodal value estimates along with the m known nodal values supplies the $\hat{\omega}(z)$ integral function with sufficient data to determine the CVBEM approximation of (9).

CVBEM APPROXIMATION ERROR

Generally, numerical approximation errors in solving potential problems is of two forms: (i) errors due to not satisfying the governing equation over Ω, and (ii) errors due to not satisfying the boundary conditions continuously on Γ. For the CVBEM, (and for other boundary integral equation methods), the first type of approximation error is eliminated due to both $\hat{\phi}$ and $\hat{\psi}$ being potential functions. But $\hat{\omega}(z)$ does not usually satisfy the boundary conditions continuously on Γ (if it did, then $\hat{\omega}(z) = \omega(z)$). The next step in the CVBEM analysis is to work with $\hat{\omega}(z)$ in order that $\hat{\omega}(z) \to \omega(z)$.

This step in the analysis of approximation error provides a significant advantage over domain numerical methods such as finite elements or finite differences. In the domain methods, the analyst examines error with a form of sequence Cauchy convergence criteria by arbitrarily increasing the domain nodal densities and comparing the resulting change in estimated nodal values. Whereas with the CVBEM, the analyst has several forms of the approximation error to work with (Hromadka, 1984b). Probably the easiest form of error to study is the development of the approximative boundary $\hat{\Gamma}$ which represents the locations where $\hat{\omega}(z)$ achieves the desired boundary values of $\omega(z)$. Generally, the boundary conditions are constant values of ϕ or ψ along boundary elements, i.e., $\phi = \phi_j$ for $z \in \Gamma_j$ or $\psi = \psi_k$, for $z \in \Gamma_k$. This set of m nodal values $\{\phi_j, \psi_k\}$ are level curves of $\omega(z)$. The approximative boundary $\hat{\Gamma}$ is determined by locating those points where $\hat{\phi} = \phi_j$ and $\hat{\psi} = \psi_k$ (see Fig. 1). Due to the collocation process, $\hat{\Gamma}$ intersects Γ at least at each nodal point location, z_j, $j = 1, 2, \cdots, m$.

To determine $\hat{\Gamma}$, each element Γ_j is further subdivided by interior points (specified by the program user) where $\hat{\omega}(z)$ is to be evaluated. At each element interior point, $\hat{\omega}(z)$ is calculated from the line integral of (9) and the values of $\hat{\phi}$ and $\hat{\psi}$ are determined. If the appropriate $\hat{\phi}$ (or $\hat{\psi}$) matches the boundary condition on Γ_j, then $\hat{\Gamma}$ intersects Γ at that point. Otherwise, subsequent points are evaluated by marching pointwise along a line perpendicular to Γ_j until the boundary condition value is reached. For point locations interior of Ω, eq. (9) is used. For points exterior of $\Omega \cup \Gamma$, an analytic continuation of (9) is used.

In this fashion, a set of points are determined where $\hat{\omega}(z)$ equals the desired ϕ_j or ψ_k values. The contour $\hat{\Gamma}$ is estimated by then connecting these points by straight lines. Because $\hat{\Gamma}$ and Γ intersect at least at nodal point locations, $\hat{\Gamma}$ appears as a plot which oscillates about the Γ contour.

COMPUTER INTERACTION FOR ERROR REDUCTION

A procedure to use a graphical display for evaluating the CVBEM model is to display both Γ and $\hat{\Gamma}$ superimposed on the CRT. By magnification of the departure between Γ and $\hat{\Gamma}$, the analyst can easily inspect the performance of the CVBEM approximation. Because the approximation error is due to the assumed basis function assumptions, the integration error is reduced by the addition of nodal points on Γ, similar to an adaptive integration technique.

The addition of nodal points can be made directly via the CRT screen and a "locating the closest boundary coordinate" computer-graphics subroutine. After the nodal additions are completed, a new $\hat{\omega}(z)$ is determined and the revised $\hat{\Gamma}$ plotted on Γ. By the addition (and deletion) of nodal points from Γ, the analyst is able to quickly evaluate the quality of the CVBEM model. Because the addition of a nodal point can be interpreted as the addition of an approximation error sink term, the geometric representation of error by means of $\hat{\Gamma}$ provides a mathematically sophisticated yet easy-to-use modeling tool.

CASE STUDY

To illustrate the previous discussion, a computer-interactive version of the CVBEM for solving potential problems in two-dimensional domains as developed by ADVANCED ENGINEERING SOFTWARE (Irvine, California) is considered.

The test problem considered is the development of a CVBEM approximation function for the two-dimensional domain shown in Fig. 4. This example represents any number of possible engineering problems such as listed in Table I.

The objective of the analysis is to locate a sufficient number of CVBEM nodal points on Γ until $\hat{\Gamma}$ is within an acceptable tolerance to Γ. Generally, this tolerance is the allowable limits of deviation from the design for construction purposes.

Using symmetry, the domain of Fig. 4 is reduced to the domain of Fig. 5. The purpose of using symmetry is to reduce computational effort and computer memory requirements. Because the CVBEM is a boundary integral method, all nodal values are linked together resulting in a square matrix. Consequently the use of symmetry to reduce the problem size, or even to use the computer-interaction approach rather than a brute force computer-generated nodal distribution on Γ, saves considerably on computational requirements.

Figure 6 shows the first attempt at modeling the domain of Fig. 5. Because of the nature of the approximative boundary concept, the boundary condition values of constant ϕ (or ψ) stepwise along Γ are of no real consequence. However, for the reader's convenience, the boundary conditions are also shown in Fig. 5.

Figure 7 shows the overlay of Γ and $\hat{\Gamma}$ for the nodal distribution used in Fig. 6. The modeler locates additional nodes for subsequent tries based on the largest departure between Γ and $\hat{\Gamma}$. After four attempts, the CVBEM modeling error is represented by $\hat{\Gamma}$ as shown in Fig. 8. It is noted that in Fig. 8, departure is magnified ten-fold for visibility. As discussed previously, if the $\hat{\Gamma}$ is acceptable for construction purposes then the associated $\hat{\omega}(z)$ is the exact solution of the boundary value problem with Γ transformed into $\hat{\Gamma}$.

SOFTWARE PACKAGE DESIGN

Both minicomputer and microcomputer versions of the discussed CVBEM technique are available. Consequently, the software structure for an Apple II E 64K microcomputer will be presented only.

The reported CVBEM computer interaction program is subdivided into 3 large legs where each leg contains the main driver program.

The program package is composed of

(i) CVBEM approximation program (to determine nodal estimates)
(ii) CVBEM approximator evaluation program (to evaluate any $\hat{\omega}(z)$)
(iii) Approximative boundary determination program to determine (x,y) coordinates where $\hat{\omega}(z)$ equals the boundary condition level curves
(iv) line drawing graphics program to plot (x,y) pairs for both Γ and $\hat{\Gamma}$ onto CRT (or plotter)
(v) Nodal point (x,y) data entry routine

The microcomputer programming is structured as shown in Fig. 9. From the figure, disc storage is used to store $\hat{\Gamma}$ related (x,y) pairs, otherwise, computer memory is used for nodal point coordinates.

IDEAL FLUID FLOW ANALYSIS

The use of the CVBEM to develop approximations of two-dimensional ideal fluid flow is documented in Hromadka (1984a). Figures 10 through 13 illustrate ideal fluid flow problems, and the approximation error developed by the CVBEM along the problem boundary.

CONCLUSIONS

The CVBEM has been used to develop highly accurate solutions for two-dimensional potential problems. In order to achieve a high degree of accuracy, a computer interactive graphics technique is reported which utilizes the approximative boundary technique to display the CVBEM modeling error as a result of the nodal point distribution selected by the analyst. Subsequent nodal point locations can be added (or deleted) by direct interaction with the computer program via the CRT. The only programming requirements needed to implement this easy-to-use analysis approach with the CVBEM is a standard CRT line-drawing graphics package, and a "locating a point to the closest contour" program routine.

REFERENCES

1. Hromadka II, T. V., "The Complex Variable Boundary Element Method," Springer-Verlag (1984a).
2. Hromadka II, T. V., "Four Methods of Locating Collocation Points for the CVBEM," Engineering Analysis (1984b).
3. Mathews, J. H., "Basic Complex Variables," Allyn & Bacon, Inc., (1982).

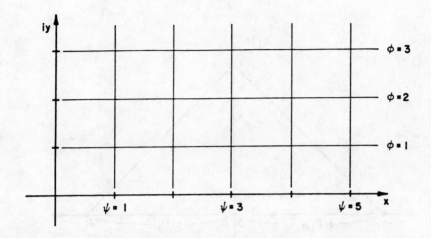

Fig. 1. Level Curves of an Analytic Function
(Example Shown: $w(z) = z$)

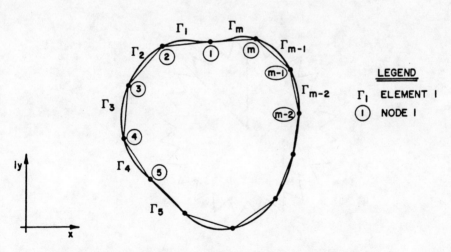

Fig. 2. Modeling Γ by Boundary Elements Γ_j

20 COMPLEX VARIABLE B E METHOD

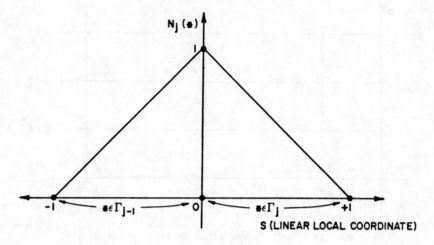

Fig. 3. Linear Basis Function

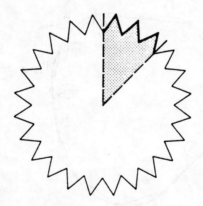

Fig. 4. Example Problem Geometry

COMPLEX VARIABLE B E METHOD 21

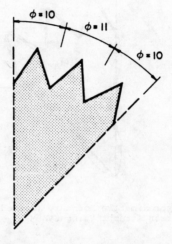

Fig. 5. Simplified Problem Geometry

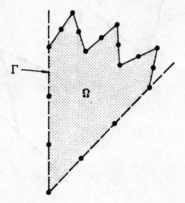

Fig. 6. CVBEM Nodal Distribution for Example Problem

22 COMPLEX VARIABLE B E METHOD

Fig. 7. Approximative Boundary (Dashed Line) for First Attempt Using CVBEM

Fig. 8. Approximative Boundary (Dashed Line) After Four Attempts Using CVBEM. (Departures between Γ and $\hat{\Gamma}$ are Magnified Tenfold)

COMPLEX VARIABLE B E METHOD 23

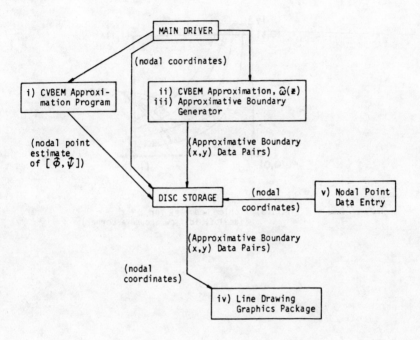

Fig. 9. CVBEM Computer-Interaction Program Structure Schematic

24 COMPLEX VARIABLE B E METHOD

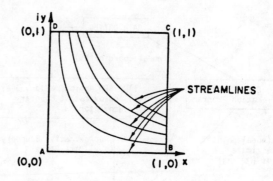

Fig. 10a. Problem Geometry for $\omega = z^2$
(ideal fluid flow around a corner)

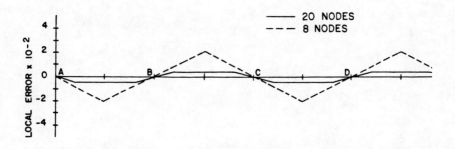

Fig. 10b. Local Error in Matching Boundary Conditions

COMPLEX VARIABLE B E METHOD 25

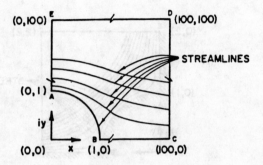

Fig. 11a. Problem Geometry for $\omega = z + z^{-1}$
(ideal fluid flow over a cylinder)

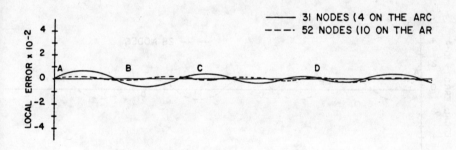

Fig. 11b. Local Error in Matching Boundary Conditions

26 COMPLEX VARIABLE B E METHOD

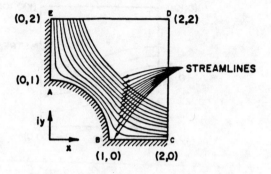

Fig. 12a. Problem Geometry for $\omega = z^2 + z^{-2}$
(ideal fluid flow around a cylindrical corner)

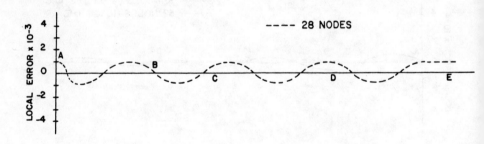

Fig. 12b. Local Error in Matching Boundary Conditions

COMPLEX VARIABLE B E METHOD 27

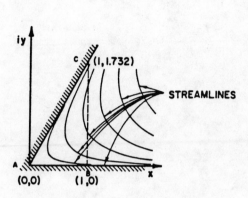

Fig. 13a. Problem Geometry for $\omega = z^3$
(ideal fluid flow around an angular region)

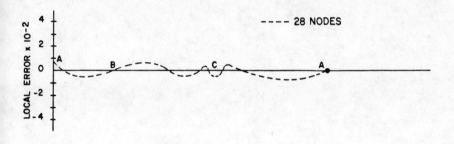

Fig. 13b. Local Error in Matching Boundary Conditions

The Solution of the Reynolds Averaged Navier-Stokes Equations in General Curvilinear Coordinates and its Application to Vehicular Aerodynamics

B.R. Patel and A.S. Dvinsky
Creare Incorporated, Hanover, New Hampshire 03755 U.S.A.

INTRODUCTION

Design of automobiles presents many challenging problems in the area of fluid dynamics. These range from the complex fluid flow phenomena occurring in the cylinders of the engine to the external aerodynamics of the vehicle. In the past, the only practical way of addressing these complex flows was to rely heavily on extensive (and expensive) cut and try experimentation. With the advent of computational fluid dynamics, the designer/engineer now has a new set of tools available that can considerably augment the experimental approach. Creare Inc., has developed two general purpose fluid flow modeling programs FLUENT and FLUENT/BFC that are finding increased use in the automobile industry. Both programs have been developed with special emphasis on ease of use and reliability. FLUENT solves the governing equations in Cartesian or cylindrical coordinate systems and offers advanced capabilities such as combustion of liquid fuels and the full Algebraic Stress Model (ASM), in addition to k-ε model for calculation in turbulent flows. FLUENT/BFC utilizes generalized curvilinear coordinates for accurate modeling in complex geometries and is the focus of this paper.

The fluid flow problems encountered in automobile design can be divided into two general categories. The first category includes problems where the flow field is governed primarily by free shear layers and mixing of fluid streams. An example is the flow field in the passenger compartment caused by the heating and air conditioning systems. In such flows, the detailed shape of the boundary has little effect on the flow field. For this class of problems, programs such as FLUENT, that solve the governing equations in a non-boundary-fitted coordinate system, are adequate. Geometric features such as curved boundaries that do not conform to the coordinate lines (surfaces) are approximated by a line (surface) natural to the used coordinate system, so a staircase-like boundary results. Although this approximation does introduce some inaccuracies in the flow prediction near the boundaries, these inaccuracies are generally tolerable for this class of flows.

The second class of flows are the wall dominated flows where the shape of the boundaries has an important effect on the flow field. This class of problems includes flow around the automobile, flows in the internal passages of the engine, and flows in the heating/cooling ducts. Given the importance of the boundaries in this class of problems, it is imperative therefore that they be represented accurately. This requires the solution of the governing fluid flow equations in generalized curvilinear coordinates. FLUENT/BFC is a general purpose fluid flow modeling program that has this capability. In the following sections, we will present a description of the techniques used in FLUENT/BFC for the solution of the governing equations in generalized curvilinear coordinates.

GOVERNING EQUATIONS

The equations used in the FLUENT/BFC 1.0 are the ensemble-averaged, steady momentum balance, mass balance, and energy balance equations:

Momentum

$$\nabla \cdot (\rho \bar{V} \bar{V}) = \nabla \cdot (\bar{\pi} + \bar{\pi}_t) \qquad (1)$$

Mass

$$\nabla \cdot \rho V = 0 \qquad (2)$$

Energy

$$\nabla \cdot (\rho \bar{V} E) = -\nabla \cdot (\bar{q} + \bar{q}_t) + \nabla \cdot \bar{V} \cdot (\bar{\pi} + \bar{\pi}_t) \qquad (3)$$

where ρ is density, $\bar{V}$ is velocity, $\bar{\pi}$ and $\bar{\pi}_t$ are the molecular and turbulent stress tensors, respectively, E is total energy, and $\bar{q}$ and $\bar{q}_t$ are the mean heat flux vector and the turbulent heat flux vector, respectively. The molecular stress tensor is defined as

$$\tilde{\pi} = 2\mu \tilde{D} - P\tilde{I} \qquad (4)$$

where $\tilde{D}$ is the deformation tensor, $\tilde{D} = \frac{1}{2}(\nabla \bar{V} + (\nabla \bar{V})^T)$, μ is molecular viscosity, and P is pressure. The turbulent stress tensor has been modeled by an isotropic eddy viscosity

$$\tilde{\pi}_t = -\rho\,\overline{\bar{v}'\bar{v}'} = 2\mu_t\,\tilde{D} - \frac{2}{3}\rho k\,\tilde{I} \tag{5}$$

where k is the turbulent kinetic energy and μ_t is the turbulent viscosity defined by an appropriate turbulence model. The mean heat flux vector and the turbulent heat flux vector are defined as follows:

$$\bar{q} = \kappa\,\nabla T \tag{6}$$

$$\bar{q}_t = \kappa_t\,\nabla T \tag{7}$$

where κ and κ_t are the mean and turbulent thermal conductivities, respectively. The total energy is defined by

$$E = c_v T + \tfrac{1}{2}\bar{V}\cdot\bar{V} \tag{8}$$

and the equation of state for a perfect gas is given by:

$$P = \rho RT \tag{9}$$

The variation of molecular viscosity and the mean thermal conductivity with the temperature is determined from Sutherland's law:

$$\frac{\mu}{\mu_0} = \left(\frac{T}{T_0}\right)^{3/2}\frac{T_0+S_1}{T+S_1} \tag{10}$$

$$\frac{\kappa}{\kappa_0} = \left(\frac{T}{T_0}\right)^{3/2}\frac{T_0+S_2}{T+S_2} \tag{11}$$

where $S_1=100$ K for air and $S_2=194$ K for air.

In present work, we have used the k-ε, two-equation turbulence model Jones and Launder[1]. In this model both the turbulence kinetic energy k and the turbulence dissipation rate are governed by the following transport equations:

$$\nabla\rho\bar{V}\varepsilon = \nabla\frac{\mu_t}{\sigma_\varepsilon}\nabla\varepsilon + C_1\,(2\mu_t\,\tilde{D}:\tilde{D})\frac{\varepsilon}{k} - C_2\rho\frac{\varepsilon^2}{k} \tag{12}$$

$$\nabla\rho\bar{V}k = \nabla\frac{\mu_t}{\sigma_k}\nabla k + 2\mu_t\,\tilde{D}:\tilde{D} - \rho\varepsilon \tag{13}$$

where the following empirical relations are used (Jones and Launder[1]):

$$C_1 = 1.43 \quad \sigma_\varepsilon = 1.3 \quad \sigma_k = 1.0$$

$$C_2 = 1.92\,[1.0 - 0.3\,\exp(-R_\tau^2)]$$

where $\quad R_\tau = \dfrac{\rho k^2}{\mu \varepsilon}$

The turbulent eddy viscosity is evaluated using the Prandtl-Kolmogorov relationship

$$\mu_t = C_\mu \frac{\rho k^2}{\varepsilon}$$

GOVERNING EQUATIONS IN COMPONENT FORM

The governing equations presented above are written in tensor form and require further transformation to become amenable to numerical solution. The first step in this direction is reformulating the equations in component form. There are generally two options which allow reduction of the tensor equations to the scalar form and still preserve the invariance under the coordinate transformation. The first is to write the governing equations in component form in any convenient coordinate system (for example, the Cartesian system) and then use chain rule differentiation to change the independent variables from original coordinates to general curvilinear coordinates while leaving the dependent variables in the original coordinate system. Another approach is to reformulate the governing equations in the contravariant form. This approach was adopted in FLUENT/BFC. Thus, equations (1) - (3) take the following form:

$$\frac{\partial}{\partial z^j}(\rho v^j v^i \sqrt{g}) + \rho v^m v^j \Gamma^i_{mj}\sqrt{g} = \frac{\partial}{\partial z^j}\sqrt{g}(\pi^{ij}+\pi_t^{ij}) + (\pi^{jm}+\pi_t^{jm})\Gamma^i_{mj}\sqrt{g} \tag{1'}$$

$$\frac{\partial}{\partial z^i}(\rho v^i \sqrt{g}) = 0 \tag{2'}$$

$$\frac{\partial \rho v^i E \sqrt{g}}{\partial z^i} = \frac{\partial}{\partial z^i}[(q^i+q_t^i)\sqrt{g}] + \frac{\partial}{\partial z^i}[g_{mj}(\pi^{ij}+\pi_t^{ij})v^m \sqrt{g}] = 0 \tag{3'}$$

where z^m, $m=1,2,3$, are general curvilinear coordinates, V^i are contravariant components of velocity vector, Γ^i_{jk} are Christoffel symbols, g_{ij} are covariant components of the metric tensor, g is determinant of the metric tensor.

The equation (1') is written in so-called semi-strong conservation form since it has terms that are not differentiated while equations (2') and (3') are written in strong conservation form. The non-conservative part of the momentum equations is due to spatially variable coordinate base vectors in general curvilinear coordinates. Otherwise, covariant derivatives of the base vectors would vanish and, hence, the Christoffel symbols would also vanish yielding a conservative form of (1'). Although the equation (1') can be transformed to the strong conservation form, the resulting equation is much more complex. In addition, it would require imposition of an additional "conservation" constraint on metric coefficients to prevent possible errors such as calculation of non-uniform flow on a non-uniform mesh while the flow is uniform in the physical problem.

The k-ε equations in component form are presented below

$$\frac{\partial \rho V^i \varepsilon \sqrt{g}}{\partial z^i} = \frac{\partial}{\partial z^i} g^{ij} \sqrt{g} \frac{\mu_t}{\sigma_\varepsilon} \frac{\partial \varepsilon}{\partial z^j} + 2 C_1 \mu_t \frac{\varepsilon}{k} D^{ij} D^{nm} g_{in} g_{jm} - C_2 \rho \frac{\varepsilon^2}{k} \quad (12')$$

$$\frac{\partial \rho V^i k \sqrt{g}}{\partial z^i} = \frac{\partial}{\partial z^i} g^{ij} \sqrt{g} \frac{\mu_t}{\sigma_k} \frac{\partial k}{\partial z} + 2 \mu_t D^{ij} D^{mn} g_{im} g_{jn} - \rho \varepsilon \quad (13')$$

GENERATION OF GRID AND DISCRETIZATION OF GOVERNING EQUATIONS

As we discussed above, the governing equations in FLUENT/BFC are formulated in contravariant form and hence they are valid in any curvilinear coordinate system. This property makes it possible to use a natural coordinate system for the problems considered. The existence of an analytically derived natural coordinate system for a given problem is an exception rather than a rule and numerical methods have been developed to facilitate the natural coordinate system generation process. The most comprehensive review of proposed methods for numerical grid generation can be found in Thompson[2] and Eiseman[3]. The procedures for grid generation can be generally divided into two categories: algebraic and differential. In algebraic methods the grid is constructed according to a set of specified rules which permit direct control over the properties of the generated grid. In addition, because of the explicit nature of the rules the grid generation process is computationally inexpensive. The main difficulty with the algebraic methods is the absence of a sufficiently general set of rules to allow grid generation in a broad class of topologies without modification of the rules.

The differential methods are based on the solution of partial differential equations with dependent variables being physical coordinates and independent variables boundary conforming (natural) coordinates. A boundary value problem can be posed for these equations by specifying correspondence between coordinates and/or their derivatives at the boundaries. The problem is then solved using finite-difference techniques and if a unique solution is obtained the coordinate transformation is defined. The control over the transformation is obtained through varying certain terms in the equations and hence it is indirect and generally less precise than in algebraic procedures. The grid generation process by differential methods is normally slower than by algebraic methods. The differential methods, however, permit a very high level of automatization and generally are more robust. The latter features are particularly important in general purpose codes like FLUENT/BFC. Therefore we selected a grid generation procedure from elliptic partial differential equations (e.g. Thompson[2]) as the standard grid generator for FLUENT/BFC. Additionally, the mechanism is provided in the code to use any user-supplied grid and/or grid generator if desired. Examples of grids generated by FLUENT/BFC are shown in Figures 1, 2 and 3.

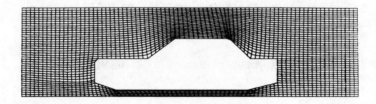

Figure 1. A fully general three-dimensional grid generation is included in FLUENT/BFC. Typical BFC geometries.

The discretized equations in FLUENT/BFC are obtained by integrating (1')-(3') around control volumes formed by the computational grid. The solution variables are defined on a staggered MAC-type grid (Harlow and Welch[4]) and the values of variables at points where they are not explicitly defined are obtained by linear interpolation. The remaining derivatives in the diffusive part of equations are approximated using central differences while convective terms are approximated using upwind differencing. A detailed description of the

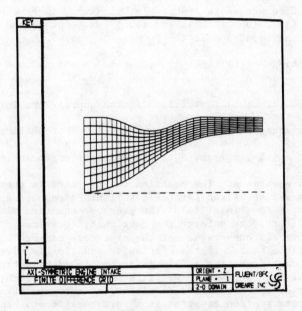

Figure 2. Example of grid generated for axisymmetric engine intake.

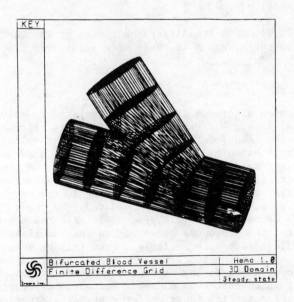

Figure 3. Example of grid generated for bifurcated blood vessel.

discretization procedure appears in Dvinsky[5]. In general equations (1'), (3'), (12'), and (13') can be presented as a set of linear equations in the form:

$$A_0 F_0 = \sum_n AF + S \qquad (14)$$

where $F=\{V^i, k, \varepsilon, E\}$, $i=1,2,3$, S is the source term in respective equations, $S=\{S_V^i, S_k, S_\varepsilon, S_E\}$. The coefficients A, $A=\{A_V^i, A_E, A_k, A_\varepsilon\}$ are obtained from corresponding

linearized equations. The equation for pressure is obtained by forming a discretized form of covariant Poisson equation from momentum equations (14). The pressure equation also serves indirectly to enforce the continuity equation. However, overall convergence of the procedure can be considerably enhanced if the continuity equation is directly solved at each iteration. This is accomplished by introducing a potential function, ϕ, as was first proposed by Pracht[6] for creeping flows. This procedure allows the exact solution of the mass conservation equation at a very small cost, because the difference operator acting on ϕ is the same as the one in the pressure equation. Using symbolic notation these two equations can now be written as

$$L(\psi) = S \qquad (15)$$

where $\psi=\{P, \psi\}$, L is the difference analog of the Laplace-Beltrami operator, and S is the source term, $S=\{S_P, S_\phi\}$.

SOLUTION ALGORITHM

The following algorithm can now be formulated to solve the system (14)-(15). First the pressure field is calculated, then momentum equations are solved to yield the velocity field. The continuity equation is solved next and the contravariant velocity field is updated. Once the pressure and velocity field are calculated the remaining variables, E, k, and ε are computed. Before proceeding to the next iteration ρ, μ, μ_t, κ, and κ_t are updated from their respective definitions. This process is repeated a number of times, until the procedure converges to the specified tolerance level for each variable. Once this happens the contravariant components of vectors are transformed to the physical components in any desired coordinate system. Presently, FLUENT/BFC uses a direct method to calculate pressure and potential function for continuity equation (Sherman[7]) the successive line overrelaxation method (e.g., Ferziger[8]) is used to calculate the remaining variables.

USER INTERFACE

Regardless of the capabilities and sophistication of the numerics, a computer program is of little value to the practising designer/engineer unless it is easy to use. Recognizing this, considerable effort has been devoted to developing the user interface for FLUENT/BFC. The developed user interface allows the user to set up the problem, run it, and analyze the obtained solution with relative ease. The inputs to the program are provided through menus and tables. The program is also designed to accept input directly from a formatted file which allows bypassing of the built-in geometric modeling capability. This feature is particularly useful when the geometry is already available from a separate CAD system.

The program has a multiple windows capability. The screen can be divided into graphics and alphanumerics windows. The user can therefore directly view the results in an interactive manner. FLUENT/BFC is designed to be device-independent, therefore the program can be used with a wide variety of graphics and alphanumeric terminals.

The geometric modeling in FLUENT/BFC is done through a CAD interface. Geometry is constructed through a hierarchy of points, curves, surfaces and unions. Points are created by specifying their coordinates. Curves are interpolated from a set of points using piecewise cubic splines. Surfaces are similarly constructed from a set of curves using bicubic surfaces. Unions are a collection of points, curves and surfaces which can be manipulated as a group. Special construction functions such as copy, arc, sweep, join etc., are provided, which considerably facilitate input of complex geometries.

A powerful three-dimensional color graphics capability has been incorporated into the program which permits viewing of all the computed flow variables. Velocity vectors, two- and three-dimensional profiles, contours and raster plots can be constructed on multiple slices throughout the solution domain. Views can be manipulated easily and the user can therefore obtain the view that most clearly depicts the flow field.

SAMPLE CALCULATIONS

A few sample calculations are presented in this section to illustrate the range of problems handled by this program. Figure 4 shows the grid for an axisymmetric duct with varying cross-section. The computed streamlines are shown in Figure 5. A prominent separated region is evident downstream of the second area contraction. The corresponding pressure contours are shown in Figure 6. In addition to these, primary, flow variables, additional, derived, parameters such as wall shear stress can be obtained. The wall shear stress for this duct is shown in Figure 7.

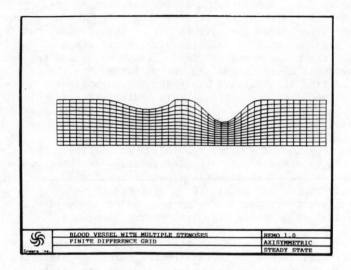

Figure 4. Finite difference grid.

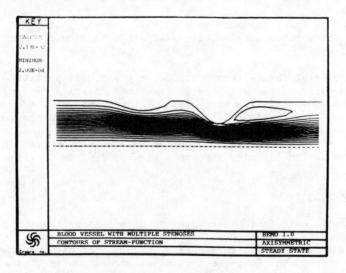

Figure 5. Contours of the stream function.

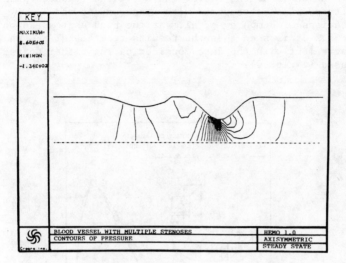

Figure 6. Contours of pressure.

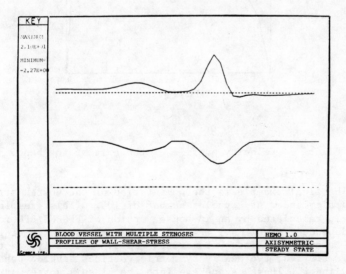

Figure 7. Wall shear stress profile.

The next example is for a flow in a rectangular cross-section 90 degree curvature duct. The computed axial velocity profile is compared with measurements and computations of Humphrey et al.[9] at the bend exit as shown in Figure 8. It is seen that the results from FLUENT/BFC show a fair agreement with the data considering the rather coarse grid used (17x16x50).

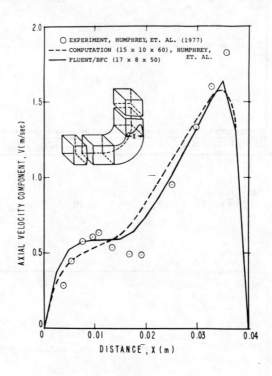

Figure 8. Velocity profile at the exit of the bend.

The final example is the computation of the vortex street behind a cylinder at Reynolds number of 100. (This example has been calculated by an in-house version of FLUENT/BFC for time-dependent problems.) Figure 9 shows the sequence of flow fields around a cylinder at various times. Initially, as the flow is started, two nearly symmetric vortices develop behind the cylinder. These degenerate into the classic vortex street as time progresses.

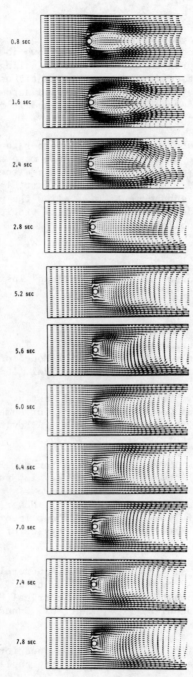

Figure 9. Vortex shedding flow past a cylinder at Re=100.

CLOSURE

A general purpose computer program FLUENT/BFC has been developed for the computation of fluid flows in curvilinear geometries. This program is capable of handling flows in many areas of automotive design. We are currently using this program to simulate the flow fields around automobiles, in various internal passages of the engine and in the heating/air conditioning ducting.

In the near future, we plan to incorporate additional capabilities in the program which include more advanced turbulence models, the ability to handle a dispersed second phase consisting of liquid droplets and particles, and combustion. These enhancements will allow the program to address an even wider range of fluid flow problems encountered in design of automobiles.

ACKNOWLEDGEMENTS

This work was supported in part by National Institute of Health Grant 1R43HL34884-01.

REFERENCES

1. Jones, W.P. and Launder, B.E. (1972), The Prediction of Laminarization with a Two-Equation Model of Turbulence; Int. J. of Heat and Mass Transfer, Vol. 15.

2. Thompson, J.E., Warsi, Z.U.A. and Mastin, C.W. (1982), Boundary-Fitted Coordinate Systems for Numerical Solution of Partial Differential Equations - A Review, J. of Computational Physics 47, pp. 1-108.

3. Eiseman, P.R. (1985), Grid Generation for Fluid Mechanics Computations; Ann. Rev. Fluid Mech., 17, pp. 487-522.

4. Harlow, F.H. and Welch, J.E. (1965), Numerical Calculation of Time-Dependent Viscous Incompressible Flow of Fluid with Free Surface, The Physics of Fluids, Vol. 8, No. 12, p. 2182-2189.

5. Dvinsky, A.S. (1987), FLUENT/BFC: A General Purpose Fluid Flow Modeling Program for all Flow Speeds, Submitted to 5th International Conference on Numerical Methods in Laminar and Turbulent Flow, Montreal, Canada, July 6-10, 1987.

6. Pracht, W.E. (1971), A Numerical Method for Calculating Transient Creep Flows, J. of Computational Physics, 7, pp. 46-60.

7. Sherman, A.H. (1981), An Empirical Investigation of Methods for Nonsymmetric Linear Systems, Elliptic Problem Solvers, Academic Press, Inc., pp. 429.

8. Ferziger, J.H. (1981), Numerical Methods for Engineering Application, John Wiley & Sons.

9. Humphrey, J.A.C., Taylor, A.M.K. and Whitelaw, J.H. (1977), Laminar Flow in a Square Duct of Strong Curvature, J. of Fluid Mechanics, V. 83, pp. 509-527.

SECTION 2 ENGINE DESIGN

Models for Predicting Engine Torque Response During Rapid Throttle Transients in Spark Ignition Engines

I.C. Finlay*, D.J. Boam*, J.J.G. Martins*, A. Gilchrist† and C.K. Lee†
*National Engineering Laboratory, East Kilbride, Glasgow
†University of Strathclyde, Glasgow

SUMMARY

Equations describing the transient behaviour of air and liquid fuel in the intake system of carburetted and single-point injected engines are used to predict firstly the variations in mixture strength and ultimately the variations in engine torque that result from rapid throttle openings. Corresponding values of measured and predicted engine torque are compared over a range of conditions. Good agreement is shown for fully warmed up conditions but with cold intake systems differences between measured and predicted values are observed. These are attributed to poor spatial or temporal distribution of liquid fuel. Finally the model is used to investigate the influence of improved fuel atomisation on levels of fuel enrichment required for best torque response.

NOMENCLATURE

A_{sm}	Inner surface area of manifold	m^2
AC_D	Effective flow area at throttle plate	m^2
a_a	Acoustic speed at ambient conditions	m/s
CR	Engine compression ratio	-
C_p	Specific heat at constant pressure	J/kg K
C_v	Specific heat at constant volume	J/kg K
D	Fuel deposition parameter	-
h	Heat-transfer coefficient	W/m^2 K
K	Air-fuel ratio (mass)	-

m	Mass	kg
$\dot{m}$	Mass flow	kg/s
N	Engine speed of rotation	r/min
n	Ratio connecting rod length/crank radius	–
Pa	Ambient air pressure	N/m^2
P'_i	Static pressure at the minimum flow area of the throttle	N/m^2
P_m	Manifold pressure	N/m^2
R	Specific gas constant	J/kg K
T	Temperature	K
t	Time	s
V	Volume	m^3
$\dot{V}_{CYL}$	Rate of change of cylinder volume	m^3/s
X	Manifold/ambient pressure ratio	–
X_1	Effective manifold/ambient pressure ratio	–

GREEK LETTERS

γ	Ratio of specific heats	–
η_{vol}	Volumetric efficiency	–
ρ_m	Density of air in manifold	kg/m^3
τ	Fuel evaporation/entrainment parameter	s

SUBSCRIPTS

a	Air
E	Engine
e	Outflow to engine
f	Fuel
i	Inflow to Manifold through throttle
L	Liquid fuel

m   Manifold air

mw   Manifold wall

v   Fuel vapour/droplets

2   Region downstream of the injection point

INTRODUCTION

Intake systems for automotive petrol engines may be very broadly classified as 'wet' or 'dry' systems. Wet systems include carburetted or single-point injected systems where the fuel is introduced at entry to the manifold and the walls are wet with gasoline. Dry systems include multi-point injected or other[1] engines where the fuel is introduced close to the inlet valve and so the manifold is largely dry. With wet systems it is observed that a substantial amount of metered fuel is deposited on the walls of the manifold to form films, pools or rivulets. The liquid fuel on the walls is slow moving relative to the air and fuel vapour. At steady-state operation evaporation, deposition and re-entrainment of droplets occurs. However, during rapid throttle openings there is net deposition of liquid fuel onto the walls. The result is that during or just following a rapid throttle opening, delivery of fuel to the engine is reduced and the mixture entering the cylinders will tend to go lean with adverse effects on engine torque response and exhaust emissions. Under these conditions fuel enrichment has to be employed to achieve acceptable driveability. With multi-point injection, if fuel is introduced close to the inlet valves then good response to throttle excursions may be achieved without the need for enrichment. This, however, allows little time for effective mixture preparation. It may therefore be advantageous, for some applications, to move the location of the injector back from the inlet valves and to employ limited enrichment to compensate for the increased fuel transit time.

With both wet and dry systems, it is clear that a proper understanding of the dynamics of the liquid fuel behaviour is required if an effective fuelling strategy is to be developed that gives good engine torque response with minimum fuel consumption and low exhaust emissions. The problem is particularly acute with lean burn engines which may require to switch rapidly from a part load mixture strength in excess of 20:1 to a wide open throttle mixture strength of 13:1. It is also central to the successful operation of three-way catalyst systems which require to control the mixture strength close to the stoichiometric to achieve high conversion efficiency at the catalyst.

The model used in this paper brings together certain aspects of various previously proposed models to provide an improved model for prediction of the transient behaviour of fuel and air in the intake system of carburetted and injected engines. Predictions from this model are compared with corresponding measured values over a range of conditions and show that the model predicts essentially the correct trends. Among the parameters investigated are manifold temperature, manifold volume and the rate of throttle opening.

MODELLING

Air flow model
When modelling the throttle manifold assembly it has invariably been assumed that the presence of fuel does not affect the air flowrate. The air flow models of Fujieda and Omaya[2] and Hires and Overington[3] allow for pulsating flow caused by piston movement during cylinder charging while those of Pearson et al[4] and Wu[5] are based on the steady-state flowrate pertaining to a particular operating condition.

It is assumed here that the air flow through the throttle is quasi-steady and that conditions in the manifold are spatially uniform at any instant.

The throttle is taken as equivalent to a convergent nozzle of variable exit area through which the air flows isentropically. The effective flow area and frictional losses are accounted for by a coefficient of discharge.

Heat transfer to the air within the manifold is accounted for using an effective heat-transfer coefficient and the temperature difference between the manifold wall temperature (constant) and the air temperature (variable). Two alternative approaches can be used when modelling the air flow, namely,

(a) the volume flowrate variation during each cylinder filling can be accounted for using details of the engine geometry and speed – pulsating flow analysis, and

(b) for a particular engine speed and engine capacity an average steady volume flowrate leaving the manifold can be used – average steady flow analysis.

It is clear that (b) above will give a smooth variation of parameters during a throttle excursion while (a) will give a periodic variation superimposed on the smooth variation obtained using (b).

Based on the above assumptions and treating the air as a perfect gas with constant specific heats the variation of flow parameters can be determined using:

(i) The mass continuity equation which involves the flowrates entering and leaving the manifold and the rate of change of mass within the manifold.

(ii) The energy equation which involves the energy flowrates entering and leaving the manifold and the rate of change of energy within the manifold.

The energy equation for the manifold is

$$\frac{d(m_m C_v T_m)}{dt} = \dot{m}_{ai} C_p T_a - \dot{m}_{ae} C_p T_m + hA_{sm}(T_{mw} - T_m).$$

Noting that

$$m_m C_v T_m = \frac{p_m V_m}{(\gamma - 1)}$$

and that V_m is the constant then this equation becomes

$$\frac{dp_m}{dt} = \frac{(\gamma - 1)}{V_m} \{\dot{m}_{ai} C_p T_a - \dot{m}_{ae} C_p T_m + hA_{sm}(T_{wm} - T_m)\}. \quad (1)$$

The state equation, since V_m is constant, gives

$$\frac{dT_m}{dt} = \frac{T_m}{p_m} \left(\frac{dp_m}{dt} - \frac{RT_m}{V_m}\frac{dm_m}{dt}\right). \quad (2)$$

For quasi-steady flow through the throttle

$$\dot{m}_{ai} = \left(\frac{dm_a}{dt}\right)_i = \frac{AC_D p_a}{R} \sqrt{\left(\frac{2C_p}{T_a}\right) \left\{1 - \left(\frac{p'_i}{p_a}\right)^{(\gamma-1)/2}\right\}^{\frac{1}{2}} \left(\frac{p'_i}{p_a}\right)^{1/\gamma}} \quad (3)$$

where p'_i is the static pressure at the minimum flow area of the throttle.

In Equation 3, since $\gamma = 1.4$,

if $\quad p_m/p_a > 0.528 \quad$ then $\quad p'_i = p_m$,

if $\quad p_m/p_a < 0.528 \quad$ then $\quad p'_i = 0.528 \, p_a$,

where 0.528 is the critical pressure ratio.

Neglecting for the moment the effect of volumetric efficiency,

$$\dot{m}_{ae} = \rho_m \dot{V}_{CYL}$$

and hence

$$\dot{m}_{ae} C_p T_m = p_m \frac{\gamma}{\gamma - 1} \dot{V}_{CYL}. \quad (4)$$

When using the pulsating flow analysis and considering a four-cylinder engine the engine geometry gives

$$\left\{\frac{dV}{dt}\right\}_{CYL} = \mathring{V}_{CYL} =$$

$$= \frac{2\pi N}{60} Ar \left[\sin\left(\frac{2\pi Nt}{60}\right) + \frac{1}{2} \frac{\sin\left(\frac{4\pi Nt}{60}\right)}{\sqrt{\left\{n^2 - \sin^2\left(\frac{2\pi Nt}{60}\right)\right\}}} \right] \quad (5)$$

which applies over each time period of $60/2N$ seconds.

For the average steady flow analysis for an engine of capacity V_E, the average 'steady' volume flowrate into the cylinder is

$$\mathring{V}_{CYL} = \frac{N}{2} \frac{V_E}{60}. \quad (6)$$

Equations 5 or 6 are substituted for $\mathring{V}_{CYL}$ in equation 4 as appropriate.

Assuming that the cylinder pressure at the end of the exhaust stroke is p_a it can be shown that the volumetric efficiency is

$$\eta_{vol} = \frac{\gamma - 1}{\gamma} + \frac{CR - p_a/p_m}{\gamma(CR - 1)}.$$

This equation indicates that η_{vol} varies with the manifold pressure ratio and, although not strictly correct for the pulsating flow analysis, equation 4 is modified as follows

$$\dot{m}_{ae} C_p T_m = p_m \frac{\gamma}{\gamma - 1} \eta_{vol} \mathring{V}_{CYL}. \quad (7)$$

Mass continuity gives

$$\frac{dm_m}{dt} = \dot{m}_{ai} - \dot{m}_{ae} \quad (8)$$

The above equations can be manipulated to give the following set of ordinary, non-linear, differential equations which constitute the air flow model.

$$\frac{dX}{dt} = \frac{1}{V_m} \left\{ \frac{\dot{m}_{ai} a_a^2}{p_a} - \gamma X \eta_{vol} \mathring{V}_{CYL} + \frac{h A_{sm}(\gamma - 1)}{p_a} (T_{mw} - T_m) \right\} \quad (9)$$

where $X = p_m/p_a$

$$\frac{dT_m}{dt} = \frac{T_m}{V_m} \left\{ \frac{a_a^2 \dot{m}_{ai}}{p_a X} \left(1 - \frac{T_m}{\gamma T_a}\right) - (\gamma - 1)\eta_{vol}\dot{V}_{CYL} \right.$$

$$\left. + \frac{hA_{sm}(\gamma - 1)}{Xp_a}(T_{mw} - T_m) \right\} \qquad (10)$$

and

$$\dot{m}_{ai} = \left(\frac{dm_a}{dt}\right)_i = \frac{AC_D p_a}{a_a} \sqrt{\left(\frac{2\gamma C_p}{R}\right) \left(1 - X_1^{\gamma-1/\gamma}\right)^{\frac{1}{2}} X_1^{1/\gamma}}. \qquad (11)$$

where $X_1 = 0.528$ if $X < 0.528$, and

$X_1 = X$ if $X > 0.528$.

The fuel flow model

The variation of mixture strength during rapid throttle transients in IC engines has been attributed to fuel evaporation delay (or transportation lag) associated with manifold wall wetting[2-4,6,7]. The fuel flow model of Fujieda and Omaya[2] arbitrarily assumed that 50 per cent of the supplied fuel was deposited as liquid on the manifold wall and was then evaporated with an assumed time delay. Hires and Overington[3] and Aquino[6] proposed fuel flow models involving liquid droplet deposition and subsequent evaporation/entrainment from the wetted manifold wall. Parameters were introduced to permit variation of the deposition and evaporation/entrainment rates. Pearson et al[4] considered the case of an initially dry manifold wall with all the liquid injected being deposited on the wall and subsequently evaporated by heat transfer at an exhaust heated hot-spot. Nishimura et al[7] did not use a fuel flow model but attempted to relate experimentally observed mixture strength excursions to engine operating conditions which could then be used as control parameters for the fuel injection rate to reduce the excursions. Wu[6] developed a computer model involving closed loop control of mixture strength during a throttle transient.

The fuel model used here follows the work of Aquino[6] and Hires and Overington[3]. Uniform conditions are assumed throughout the manifold (or individual manifold sub-region) at any instant. A fraction D of the injected fuel is assumed to be deposited onto the wall and the mass of fuel evaporated and entrained from the film is assumed to be directly proportional to the mass of liquid present in the film. The deposition parameter D and the proportionality constant for evaporation/entrainment, τ are assumed constant during a throttle transient. Computer solution of the complete model, air and fuel flow, permits investigation of the effects of the parameters D and τ. Figure 1a illustrates the model for the case of single-point injection and Figure 1b is used when considering multi-point injection.

The mass flowrate of fuel injection is expressed as $\dot{m}_{fi} = \dot{m}_{ai}/K$ where K can be constant, at say a value giving the correct air/fuel ratio for steady-state operation, or can be varied with time in a chosen manner in order to limit the magnitude of the mixture strength variation.

It is assumed that the air/fuel ratio leaving the manifold is equal to the ratio mass of air/mass of fuel outwith the liquid film, which exists within the manifold volume.

For single-point injection mass continuity of the fuel flow gives

$$\frac{dm_L}{dt} = D \frac{\dot{m}_{ai}}{K} - \frac{m_L}{\tau} \tag{12}$$

and

$$\frac{dm_v}{dt} = \frac{\dot{m}_{ai}}{K} - \frac{dm_L}{dt} - \dot{m}_{ae}\left(\frac{m_v}{m_a}\right). \tag{13}$$

The air/fuel ratio entering the engine at any instant is given by the ratio m_a/m_v.

For multi-point injection a film is considered to be present only in the region downstream of the injection point. Thus the manifold can be subdivided into two regions. Mass continuity for the fuel gives the following

$$\frac{dm_L}{dt} = D \frac{\dot{m}_{ai}}{K} - \frac{m_L}{\tau} - \dot{m}_{Le} \tag{14}$$

where $\dot{m}_{Le}$ can be calculated from the shear stresses on the boundary between gas and liquid phases

and

$$\frac{dm_v}{dt} = \frac{\dot{m}_{ai}}{K} - \frac{dm_L}{dt} - \dot{m}_{ae}\left(\frac{m_v}{m_a}\right)_2. \tag{15}$$

Solutions can be obtained for the mixture strength variation by simultaneous solution of the air and appropriate fuel model. It should be noted that the air flow model is identical for both fuel flow models being considered.

Solutions obtained for a particular case using the average steady flow model and then using the pulsating flow model confirmed that use of the latter model simply produces a high frequency parameter variation about the average parameter variation obtained using the former model.

TEST RIG

The test rig on which the the reported measurements were made is described in detail in Reference 8. The engine used was a 1.6 litre, four-cylinder unit fuelled by a throttle body injector. As shown on Figure 2 the driveshaft of the engine was coupled to a set of flywheels and to an eddy current dynamometer. The flywheels were sized to simulate a typical 1600 cc saloon in top gear. To enable fast and reproducible throttle excursions to be imposed on the engine the throttle plate was actuated by a stepping motor controlled by an HP85 computer. The steady state and transient fuelling was also controlled from an HP85 unit. The torque response of the engine was determined by accurately measuring flywheel velocity and differentiating to determine flywheel acceleration and hence engine torque. Variations in air/fuel ratio were recorded using a Cusson's Lamdascan unit but the slow response of this instrument made it unsuitable for use during fast transients.

RESULTS

Figure 3 shows the torque response obtained from the engine when the throttle was taken from a road load setting at 1570 r/min in top gear to the wide open position (90°) in 0.3 s. The engine was fully warmed-up and the intake manifold heated by engine coolant at 90°C but no transient enrichment was used. As may be seen from Figure 3(a and b) the fuelling was simply switched from the road load level to the WOT level the instant the throttle started to move. This fuelling strategy clearly gives a very unsatisfactory torque response as shown in Figure 3c. Immediately following the start of the throttle opening the torque rises but then collapses completely for approximately four engine revolutions before recovering strongly to the WOT level over the next six engine revolutions.

Although no transient enrichment was employed it is clear from Figure 3(b and d) that during the first two engine cycles immediately following the start of the transient the sudden switch to WOT fuelling level occurs before the throttle has opened sufficiently to allow maximum air flow to the engine. No measurements of the instantaneous air flowrate were made but data predicted by Equations 8-11 are shown in Figure 3d and these confirm that it takes almost two engine revolutions for the air flowrate to reach the WOT value. The variation with time of the mixture strength arriving at the intake valves of the engine as predicted by Equations 12 and 13 is shown in Figure 3(e and f). Clearly after an initial enrichment period of 1-2 engine cycles the mixture strength quickly leans off beyond the limit of stable combustion and takes several engine cycles to recover. The speed at which the mixture strength leans off is strongly influenced by the deposition factor D and the rate at which the mixture strength recovers is determined by the proportionality factor $1/\tau$. The influence of D and τ on the predicted excursions in mixture strength are shown in Figure 3(e and f).

As stated previously it was not possible to measure the excursions in mixture strength with sufficient accuracy owing to the relatively slow response of the available equipment. Instead, therefore, the predicted excursions in air/fuel ratio were converted to engine torque using measured values of the relationship between engine torque and air/fuel ratio an example of which is shown in Figure 4. The predicted torque values obtained by this method are shown in Figure 5 alongside corresponding measured values. Manipulation of τ and D to give a good fit to the measured data gave a value of D = 0.84 and τ = 0.78. D = 0.84 suggests that 16 per cent of the fuel leaving the injector moves directly into the engine as vapour or entrained droplets and that the remaining 84 per cent spends a significant amount of time on the walls in the liquid phase. This is in broad agreement with earlier studies[9][10] which showed that between 15 and 20 per cent of the gasoline leaving the jets of a carburettor flashes off as vapour before reaching the plenum of the manifold. A τ value of 0.78 indicates that it takes approximately 1.8 s for the remainder 90 per cent of the fuel to reach the engine either as vapour or as liquid droplets.

As may be expected, adding enrichment fuel during the throttle opening period improves engine response. Figures 6 and 7 show the effects of adding increased amounts of enrichment fuel at the same initial engine conditions and rate of throttle opening as before. The initial rises in torque are now sustained for larger periods and the subsequent dips progressively reduced. Manipulating D and τ to obtain predicted values close to the measured data again gave values of 0.84 and 0.78 respectively.

As the temperature of the intake system is lowered so the rate of evaporation should reduce but the deposition level should remain unchanged. τ should therefore increase and D remain at 0.84. Figure 8 shows a comparison of measured and predicted response data when the engine is subjected to a fast (0.3 s) throttle opening with the enrichment levels of Figure 7 but with the intake manifold temperature reduced to 30°C. Important differences are now apparent between measured and predicted values. The initial rise in torque is accurately predicted as is the subsequent fall but the recovery is poorly predicted.

This may be due to two-phase flow effects, not allowed for in the model, becoming significant at this condition. When the manifold is at a temperature of 30°C the reduced fuel evaporation rate will result in increased amounts of liquid fuel being present on the walls of the intake system. This could cause a deterioration in the uniformity of both the spatial and temporal supply of fuel to the cylinders. The positive level of measured torque generation seen on Figure 8 may therefore have been produced by either intermittent firing of all cylinders or continuous firing of some cylinders. A model that assumed perfect distribution both spatially and temporally may therefore have all cylinders in the misfire region and so would predict low torque as shown in Figure 8. As more enrichment fuel is provided so the torque response improved as shown in Figures 9 and 10. Figure 10 shows the torque response of the engine when fuelling is close to optimal. At 30°C a value of $\tau = 2.5$ gives predicted responses that are close to the measured values as shown in Figures 8-10.

It is clear from the foregoing discussion that the need for enrichment fuelling stems from the hold-up of liquid fuel on the surface of the manifold. Reducing hold-up of liquid fuel by increasing the temperature of the manifold, ie reducing τ, is unacceptable because of the adverse effects on engine knock resistance and power. Reducing the quantity of liquid fuel impacting on the walls of the manifold, ie reducing D by, for example, air assisted atomisation[11] or ultrasonic atomisation[12] may help to reduce the level of fuel enrichment required. Much of the work on the influence of improved fuel atomisation on engine performance has been carried out under steady-state conditions and has concentrated on effects such as intercylinder distribution, emissions etc. It is worth noting that improving fuel atomisation may also improve fuel economy by reducing the level of enrichment required during transients. Figure 11 quantifies this by showing levels of enrichment required when D is reduced from 0.8 to 0.6. A change in D from 0.8 to 0.6 reduces the fuel enrichment level required by approximately 50 per cent. As before the transient condition studied was that of moving from 1570 r/min road load in fourth gear to wide open throttle in 0.3 s with a manifold temperature of 90°C.

CONCLUSIONS

1   Measured and predicted torque responses obtained from a single-point injected, multicylinder, gasoline engine are compared over a range of conditions.

2   When the intake system is fully warmed-up agreement between measured and predicted values is shown to be good. Even when inadequate levels of enrichment are used and the torque response is poor, predicted values closely follow the measured response.

3   When the temperature of the manifold is reduced to 30°C, some differences between measured and predicted values are found. These differences are attributed to poor spatial or

temporal distribution of liquid fuel. Agreement between measured and predicted values appears to improve as the enrichment levels used approach the optimum.

4 A value of D = 0.84 appears to give best agreement between measured and predicted values at all conditions. This value agrees with earlier studies. τ is found to vary from 0.78 with a hot (90°C) manifold to 2.5 when the manifold is at 30°C.

5 The model is used to predict the effect of improved fuel atomisation on the levels of enrichment fuelling required. Reducing D from 0.84 to 0.6 is estimated to reduce the enrichment fuelling level by up to 50 per cent.

REFERENCES

1. Emmenthal K.D. et al (1985), Air Forced Injection System for Spark Ignition Engines. SAE Paper No 850483, Society of Automotive Engineers.

2. Fujieda M. and Oyama, Y. (1984), Analysis of Transient Mixture Transport in Intake Manifold for Carburetted Engine. J.S.A.E. Review.

3. Hires S.D. and Overington M.T. (1981), Transient Mixture Strength Excursions - An Investigation of Their Causes and the Development of a Constant Mixture Strength Fuelling Strategy. SAE Technical Series, 810495, International Congress and Exposition, Detroit, Michigan.

4. Pearson J.K., Orman P.L. and Caddock B.D. (1983), Car Driveability Modelling: A Computer Model for the Prediction of Hesitation Under Cold Weather Accelerating Conditions. Automotive Engineer.

5. Wu H. (1980), A Computer Model for a Centrally-located, Closed-loop, Automotive Fuel Metering System. ASME International Computer Technology Conference, San Francisco, CA.

6. Aquino C.F. (1981), Transient A/F Control Characteristics of the 5-litre Central Fuel Injection Engine. SAE Technical Paper Series, 810494, International Congress and Exposition, Detroit, Michigan.

7. Nishimura Y., Oyama Y. and Sasayama T. (1981), Transient Response of Fuel Supply Systems for Carburettor Engine. SAE Technical paper Series, 810788, Passenger Car Meeting, Dearborn, Michigan.

8. Boam D.J., Finlay I.C. and Fairhead G. (1987), The Optimisation of Fuel Enrichment Patterns During Engine Power Transients. Paper to be presented to IMechE/SAE VECON '87 Conference, September 1987.

9. Finlay I.C., Boam D.J. and Bannell J.L.K. (1979), Computer Model of Fuel Evaporation in Air Valve Carburettors. Automotive Engineer, December 1979.

10. Boam D.J. and Finlay I.C. (1979), A Computer Model of Fuel Evaporation in the Intake of a Carburetted Petrol Engine. I.Mech.E., Paper C89/79, June 1979.

11. Uozumi J. and Iwata K. (1986), Fundamental Study on Steady Characteristics of an Air Fuel Injector. J.S.A.E. Review, Vol. 7, No 2, July 1986.

12. Droughton J.V. et al (1984), The Effect of Ultrasonic Mixing of Fuel and Air on the Performance of an Internal Combustion Engine. SAE Paper No 840238.

60 PREDICTING ENGINE TORQUE RESPONSE

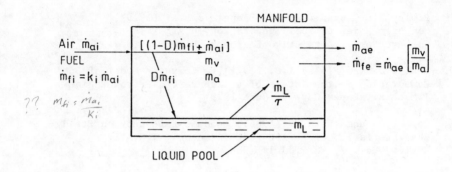

(a) Single-point Injection

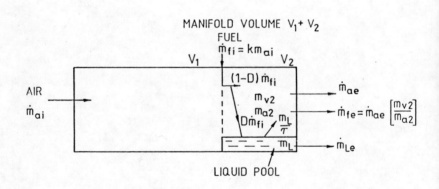

(b) Multi-point Injection

FIG 1 FUEL FLOW MODELS

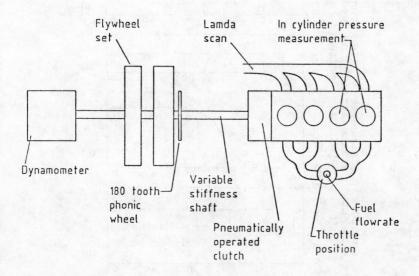

FIG 2   LAYOUT OF TEST RIG SHOWING INSTRUMENTATION USED

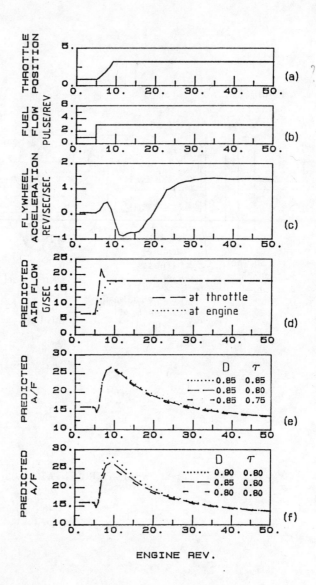

FIG 3 ENGINE RESPONSE TO A FAST THROTTLE OPENING: COMPARISON OF MEASURED ACCELERATION AND PREDICTED AIR FUEL RATIO EXCURSION

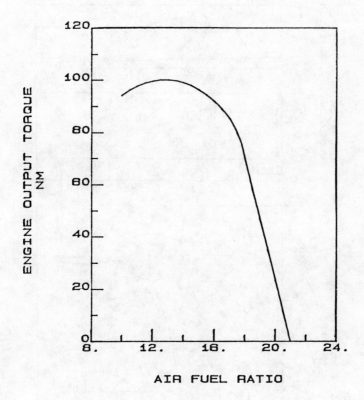

FIG 4 VARIATION OF ENGINE OUTPUT TORQUE WITH AIR FUEL RATIO AT 2000 r/min WIDE OPEN THROTTLE

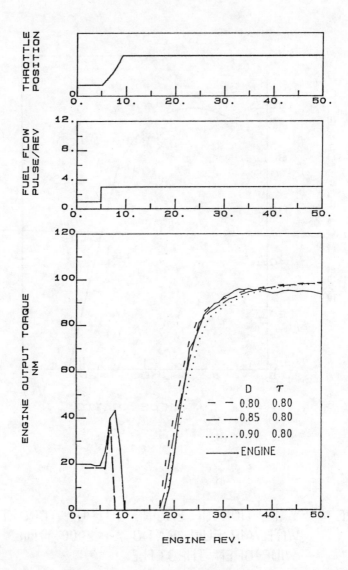

FIG 5(a) EFFECT OF D ON TORQUE PREDICTION
90°C, FAST RAMP, NO ENRICHMENT

PREDICTING ENGINE TORQUE RESPONSE 65

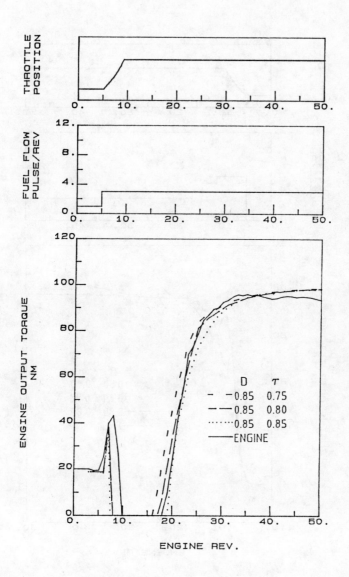

FIG 5(b) EFFECT OF τ ON TORQUE PREDICTION
90°C, FAST RAMP, NO ENRICHMENT

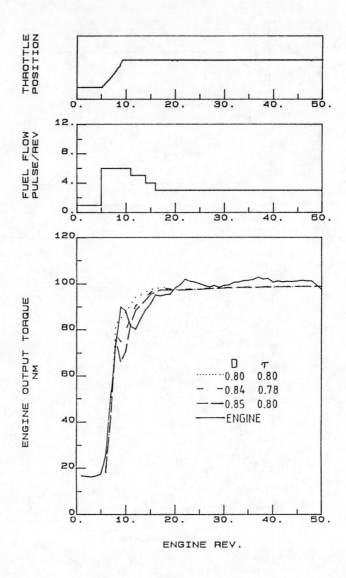

FIG 6 PREDICTED RESPONSE WITH PART ENRICHMENT 90°C, FAST RAMP

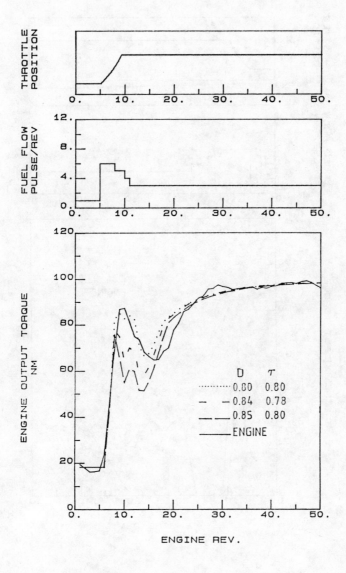

FIG 7 PREDICTED RESPONSE WITH NEAR OPTIMUM ENRICHMENT. 90°C, FAST RAMP

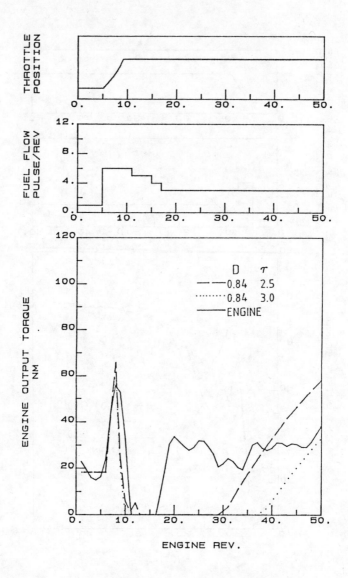

FIG 8 PREDICTED RESPONSE AT 30°C WITH FUELLING NEAR OPTIMUM FOR 90°C, FAST RAMP

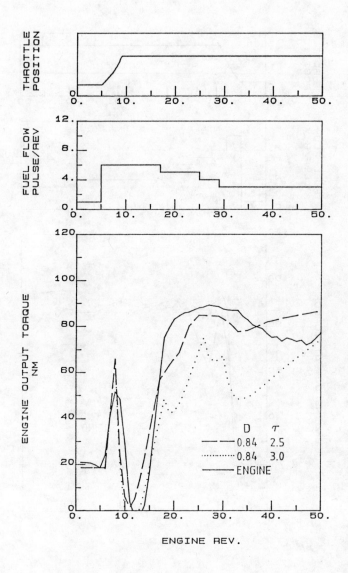

FIG 9 PREDICTED RESPONSE WITH INCREASED ENRICHMENT
30°C, FAST RAMP

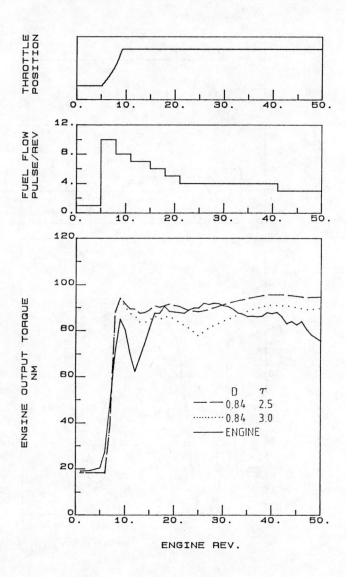

FIG 10 PREDICTED RESPONSE WITH NEAR OPTIMUM ENRICHMENT. 30°C, FAST RAMP

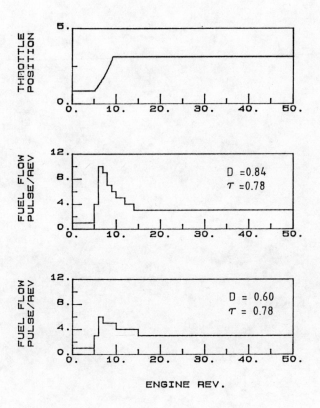

FIG 11 EFFECT OF D ON PREDICTED FUEL ENRICHMENT FOR OPTIMUM TORQUE RESPONSE TO A FAST TRANSIENT

An Optimization of Characteristic Parameters of Triple Stage Non Linear Torsional Damper in Automotive Clutch

Shao Cheng and Lü Zhenhua
Test Centre, Jilin University of Technology, Changchun, Jilin Province, People's Republic of China

INTRODUCTION

Applying torsional damper in dry clutch is the main structural device to reduce the torsional vibration and shock and the noise occurring in automotive drive train when the automobile is either moving or idling; nonlinear torsional multiple stage damper is particularly effective. Because triple stage damper is now a type widely used, this paper tries to give the optimization method and procedure of its performance parameters.

Here we consider a kind of nonlinear torsional damper which has three stages in the forward direction but two stages in the backward direction; its elastic characteristic is shown in Figure 1. The first stage is applied to reducing the idling torsional vibration and noise occurring in clutch-gearbox system. The second and the third ones are applied to reducing the torsional vibration and noise and to absorbing the torsional shock occurring in drive train when automobile is moving. Because the function of every stage of the damper is different, the mathematical model for the optimization of every stage must be different from the others. So the performance parameters of every stage of the damper should be optimized separately. In this paper, we only study the optimization of the second and the third stages. The optimization will take the effective reduction of torsional vibration response of automotive drive train as its objective. So this is an optimization problem based on dynamic response of system.

In Figure 2, the nonlinear dry friction damping characteristic of the torsional damper is shown.

MATHEMATICAL MODEL FOR THE OPTIMIZATION

A reasonable mathematical model for optimization of the torsional damper can be constituted by selecting appropriate design variables, by constructing proper objective function, and by appending necessary constraint conditions.

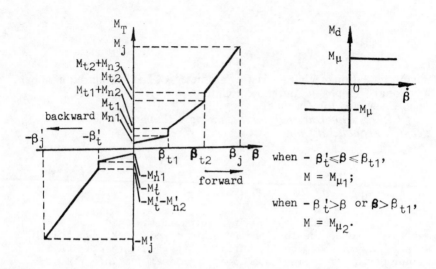

Figure 1. Elastic Characteristic of the Triple Stage Damper.

Figure 2. Damping Characteristic of the Damper.

Design variables of the damper should include structural parameters, such as medium diameter, material diameter, effective circle number of working springs in the damper, and dimensions of working windows in which the springs are enclosed; they also should include performance parameters, namely, those shown in Figures 1 and 2: torsional limit angle β_j, limit torque M_j, every critical point's parameters β_t and M_t, every stage's preacted torque M_n, and dry friction damping torque $M\mu$. The process of the optimization is to determine appropriate structural parameters so as to have the combination of the performance parameters become an optimal project, which can most effectively reduce the torsional vibration and shock, and the noise occurring in automotive drive train.

Determination of the structural parameters is related to the material's mechanical properties, the process technology of the parts, the spring design norms, and other complex factors. Therefore, here we first take the performance parameters as independent design variables. This not only will be the basis of further optimization of the structural parameters but also can be applied independently.

PARAMETERS OF NON LINEAR TORSIONAL DAMPER 75

According to preceding section, the parameters β_{t1}, M_{t1}, M_{n1}, β_t', M_t', M_{n1}' of the first stage of the damper can be given as known ones; their values may be chosen as reasonably as possible from experience.

Once the structural type and principal dimensions of the clutch are given, the available maximum value of β_j, in general, is also fixed. Therefore it need not be taken as a design variable.

So, in the elastic characteristic of the damper, the parameters needing to be optimized are M_j, β_{t2}, M_{t2}, M_{n2}, M_{n3}. As to M_{n2}', it is found as

$$M_{n2}' \approx M_{n2} + M_{n3} \tag{1}$$

and other parameters, such as k_{t1}, k_{t2}, k_{t3}, k_{t1}', k_{t2}', and M_j', are all not independent, and they can be expressed by preceding parameters; i.e.,

$$\begin{cases} k_{t1} = (M_{t1}-M_{n1})/\beta_{t1} \\ k_{t2} = (M_{t2}-M_{t1}-M_{n2})/(\beta_{t2}-\beta_{t1}) \\ k_{t3} = (M_j-M_{t2}-M_{n3})/(\beta_j-\beta_{t2}) \end{cases} \tag{2}$$

$$\begin{cases} k_{t1}' = (M_t'-M_{n1}')/\beta_{t1}' \\ k_{t2}' \approx k_{t3} \end{cases} \tag{3}$$

$$M_j' = M_t' + M_{n2}' + k_{t2}'(\beta_j'-\beta_t') \tag{4}$$

The dry friction damping torque $M\mu$ is also an important performance parameter to be optimized. However, the optimal value of $M\mu$ for the idling stage of the damper is generally not the same as that for the driving stages, which are denoted separately by $M_{\mu 1}$ and $M_{\mu 2}$. From previous explanation, value of $M_{\mu 1}$ may be given in advance, but $M_{\mu 2}$ is taken as a design variable.

According to above analyses, the design variables for the driving stages' performance parameters of the triple stage damper to be optimized are:

$$X = \{x_1, x_2, x_3, x_4, x_5, x_6\}^T$$
$$= \{M_j, \beta_{t2}, M_{t2}, M_{n2}, M_{n3}, M_{\mu 2}\}^T \tag{5}$$

where the value ranges of M_{n2} and M_{n3} are quite narrow, so after obtaining some design experiences, we may not take them as design variables.

Whether the performance parameters of the torsional damper are optimal or not is indicated by the torsional vibration response magnitude of automotive drive train and its frequency characteristic. The former is expressed by torsional angle displacement φ, angle velocity $\dot{\varphi}$, angle acceleration $\ddot{\varphi}$, etc.; the latter is shown by seeing whether there are the main critical speed points under the frequently used driving conditions of the automobile in the clutch of which a torsional damper is adopted. These two varieties of characteristics may be expressed as either objective functions or constraint conditions for the optimization. Thus there is not only one way to construct objective function. In this presentation, we take the function of maximum values of φ, $\dot{\varphi}$, and $\ddot{\varphi}$ as the objective function to be minimized for optimal design of the damper; i.e.,

$$F(X) = W_d \sum_{i=1}^{n} W_{di} f_d(|\varphi_i(x, t)|_{max})$$

$$+ W_v \sum_{i=1}^{n} W_{vi} f_v(|\dot{\varphi}_i(x, t)|_{max})$$

$$+ W_a \sum_{i=1}^{n} W_{ai} f_a(|\ddot{\varphi}_i(x, t)|_{max}) \qquad (6)$$

where f_d, f_v, and f_a may be the mean value functions or the squared-sum functions or the root-mean-squared value functions or other function forms. The subscript i stands for the sequential number of degrees-of-freedom of the torsional vibration analysis model of automotive power train. The coefficients W_d, W_v, and W_a are used to adjust the value's differences in magnitude among φ, $\dot{\varphi}$, and $\ddot{\varphi}$; the coefficients W_{di}, W_{vi}, and W_{ai} are used to reflect the relative importance of every component of $\{\varphi_i\}$, $\{\dot{\varphi}_i\}$, and $\{\ddot{\varphi}_i\}$. These coefficients may be changed in a way in the optimization computation process, but in most cases, they may be constant quantities.

Obviously, here we employ the method of combination with weights to change a multiple objective optimization problem into a single objective one. The values of all divisional objective functions f_d, f_v, and f_a, which all are implicit functions of design variable vector X, can be obtained by the simulant computation of forced nonlinear torsional vibration of automotive power train. The Equation 6 contains 3n divisional objective functions, but sometimes they may not all be required. On the other hand, we may establish some new divisional objectives according to some standard indices about the vehicle vibration and noise.

Then we can propose the following constraint conditions to the design variables of Equation 5:

1) We may limit the relative torsional angle displacements between adjacent parts of automotive drive train on the basis of the system's torsional vibration norms or design and research experiences; i.e.,

$$|\varphi_i - \varphi_{i+1}| \leq \Delta\varphi_{iu}, \quad i=1,2,\cdots, n-1 \tag{7}$$

They are implicit constraints to the design variable vector X.

2) According to the former research results on the torsional vibration characteristic of automotive power trains, we must constrain the super bound of the system's third natural frequency f_{n3}:

$$f_{n3} \leq f_{n3u} \tag{8}$$

This is also an implicit constraint, where the value of f_{n3} may be obtained by analysing the system's natural torsional vibration characteristic.

3) Considering practicable value range of every stage's stiffness of the damper, we may specify the following:

$$\begin{cases} k_{t21} \leq k_{t2} \leq k_{t2u} \\ k_{t31} \leq k_{t3} \leq k_{t3u} \end{cases} \tag{9}$$

and

$$k_{t3} > k_{t2} > k_{t1} \tag{10}$$

They are explicit constraints to X.

4) The essential constraints to design variables are to provide in advance the upper and lower bounds of every variable's value lest they should be given some insignificant values to waste computer time and even to result in computing process going astray in the optimization iteration procedure; i.e.,

$$x_{i1} \leq x_i \leq x_{iu}, \quad i=1, 2, \cdots, n \tag{11}$$

To sum up, Equations 5 to 11 constitute the mathematical model for the dynamic response optimization of the partial performance parameters of the nonlinear triple stage damper.

COMPUTATIONAL METHOD FOR THE OPTIMIZATION

The above mentioned mathematical model is a constrained nonlinear mathematical programming problem. However, the objective function F(X) and the constraint Equations 7 and 8 can not be written as explicit expressions of the design variables. So in order to compute those functions' values, we have to make the simulant computation of the nonlinear torsional vibration response and the natural frequencies of automotive power train. In optimization iteration process, for every step's feasible design project the objective and constraint functions' values are computed; thus the reanalysis of the system's dynamic response and natural frequencies must be repeatedly carried out, and the great majority of CPU time is used to make the reanalyses in the computational process; that is one of the features of dynamic response optimization of a vibration system.

As micro- and mini-computers are widely used, whether an expected solution can be obtained in a short CPU time becomes a key point of a successful optimization. We may seek for the way to shorten the time in two directions: one is to develop proper computational methods, which may have a higher computational efficiency or may reduce the computational amount, for the system's forced nonlinear torsional vibration simulation and for approximate reanalysis of the system's natural frequencies; another is to select an appropriate optimization algorithm which may reduce the number of times of iterative computations or the number of times of computing the implicit objective and constraint functions' values. For the former we shall specially discuss the questions in another paper. Here we only provide some of our understanding and considerations for the latter.

As we know, the algorithms for both constrained and unconstrained nonlinear programming can be divided into two categories: one which requires to compute the gradient of objective function, and some algorithms of which further need to determine the second partial derivatives of objective function; the other which does not require to determine the derivatives. Using the partial derivatives of $F(X)$ with respect to X may generally improve the effectiveness of every step in iterative computation and hence may reduce the total iteration number of times. However, because $F(X)$ is not written as the explicit form of X, its partial derivatives have to be approximately replaced by its difference quotients. If the gradient $\frac{\partial F(X)}{\partial X}$ is determined one time by means of the Forward Difference Quotient Method or Two-Point Central Difference Quotient Method, NV+1 or 2NV+1 values of $F(X)$ are respectively needed to be computed (NV stands for the number of design variables). If in order to improve the accuracy, it may be necessary to employ the Four-Point Difference Quotient Method to determine $\frac{\partial F(X)}{\partial X}$, or perhaps it is required to find the second partial derivative matrix (namely, Hesse Matrix) of $F(X)$, the number of times of computing the value of $F(X)$ must be too many to be practicable. In that case we have to give up the programmig algorithms of this category and apply the methods of the other category without partial derivatives of $F(X)$.

PARAMETERS OF NON LINEAR TORSIONAL DAMPER   79

In both direct and indirect solving methods for constrained nonlinear programming problem, there are the algorithms that meet above requirements. What belongs to the former includes the Net Method, the Random Test Method, the Complex Method, etc.; what belongs to the latter includes the Multiplier Method and the Penalty Function Methods, in which the unconstrained nonlinear programming algorithms without the partial derivatives of the penalty function, such as the Rotating Direction Method, the Pattern Search Method, the Powell's Method, the Simplex Method, etc., are employed. The direct algorithms are relatively simple to programme and easy to understand, and they have no special demands on objective and constraint functions; however, they will result in a very great amount of computation and lower computational efficiency, and they are disadvantageous to improve the optimization precision. Therefore, we adopt an indirect algorithm, i.e., the Sequential Unconstrained Minimization Technique (namely, SUMT for short), which is based on the Mixed Penalty Function Method, and in which the Powell's Method is employed to make the unconstrained minimization.

When we apply the SUMT to solving following standard constrained nonlinear optimization problem

$$\min F(X) \tag{12}$$

$$\begin{cases} g_i(X) \geqslant 0, & i=1, 2, \cdots, l \\ h_j(X) = 0 & j=1, 2, \cdots, m \end{cases} \tag{13}$$

the Interior Penalty Method is employed in constituting the penalty term of inequality constraints $\{g_i(X)\}$, and the Exterior Penalty Method is employed on equality constraints $\{h_j(X)\}$; the penalty function is expressed as

$$P(X, r) = F(X) + r \sum_{i=1}^{l} \frac{1}{g_i(X)} + \frac{1}{\sqrt{r}} \sum_{j=1}^{m} [h_j(X)]^2 \tag{14}$$

The penalty factor r is given values as a decreasing positive number sequence in the course of optimization iteration, and when the value of r is changed one time, a new penalty function P(X, r) is constructed.

When $r = r^{(k)}$, $P(X, r^{(k)})$ is minimized with the Powell's Method, and the corresponding minimal point $X^*(r^{(k)})$ is found. For a sequence $\{r^{(k)}\}$, when $k \to \infty$ and $r^{(k)} \to 0$, the minimal point sequence $\{X^*(r^{(k)})\}$ will lead to:

$$\begin{cases} \lim_{k \to \infty} \{r^{(k)} \sum_{i=1}^{l} 1/g_i(X^*(r^{(k)}))\} = 0 \end{cases}$$

$$\left\{ \lim_{k \to \infty} \frac{1}{\sqrt{r^{(k)}}} \sum_{j=1}^{m} [h_j(X^*(r^{(k)}))]^2 = 0 \right. \tag{15}$$

That hence yields

$$\begin{cases} \lim_{k \to \infty} \{X^*(r^{(k)})\} = X^* \\ \lim_{k \to \infty} \{P(X^*(r^{(k)}))\} = F(X^*) \end{cases} \tag{16}$$

where X^* and $F(X^*)$ are just the right solution of original constrained minimization problem. In practical solving process, when k is great enough and $r^{(k)}$ is little enough, the $X^*(r^{(k)})$ and $F(X^*(r^{(k)}))$ may be taken, at the given precision, as an approximate optimal solution of the original problem. So this algorithm prescribes the convergence criterion of the unconstrained minimization as

$$\| X^*(r^{(k)}) - X^*(r^{(k-1)}) \| \leq \varepsilon_1 \tag{17}$$

and that of whole SUMT process as

$$\left| \frac{P(X^*(r^{(k)})) - P(X^*(r^{(k-1)}))}{P(X^*(r^{(k-1)}))} \right| \leq \varepsilon_2 \tag{18}$$

where ε_1 and ε_2 are error bounds specified beforehand.

According to what has been stated above, it can be seen that in order to solve a constrained nonlinear optimization problem with the SUMT, we actually have to solve a series of unconstrained nonlinear minimization problems; thus the computational amount is quite great. Some practical computations show that in all the minimization process several hundreds of values of objective function are computed.

The one-dimensional search method of the above algorithm is that the search interval of optimal step length is first determined with the Extrapolation Method, and then the optimal step length is found with the Second Order Interpolation Method. That is a efficient acceleration technique.

Initial vector $X^{(0)}$ is demanded to be a feasible point (namely, it must meet all the inequality constraints). Otherwise a feasible initial point will be produced by means of a pseudorandom number generation subroutine; therefore the upper and lower bounds of X should be given in advance.

Given initial penalty factor value $r^{(0)}$, the sequence $\{r^{(k)}\}$ is produced as

$$r^{(k)} = C \cdot r^{(k-1)} \tag{19}$$

where $C(<1.0)$ is also a beforehand given number. The values of $r^{(0)}$ and C have much to do with the convergence or the divergence of the iterative process and with its convergence rate. However there is not a dependable principle or method for selecting those factors' values as yet, and for each problem some trial computations therefore have to be made for the selection; that is one of the present drawbacks of the Penalty Function Method. If the initial value of r is not input, the programme can determine $r^{(0)}$ as

$$r^{(0)} = \left| F(X^{(0)}) / \sum_{i=1}^{l} \frac{1}{g_i(X^{(0)})} / \sum_{j=1}^{m} h_j(X^{(0)}) \right|, \qquad (20)$$

but if there is not equality constraint, the above formula is shortened as

$$r^{(0)} = \left| F(X^{(0)}) / \sum_{i=1}^{l} \frac{1}{g_i(X^{(0)})} \right| \qquad (21)$$

A PRACTICALLY NUMERICAL EXAMPLE

Now applying the constituted mathematical model and the selected computational algorithm, we shall optimize a triple stage damper's design project of the truck NJD-131. However according to the features of this specific problem, we may make some simplifications or supplementations to the mathematical model. This is also what every designer, who wants to use a current method for optimizing a project, will firstly do.

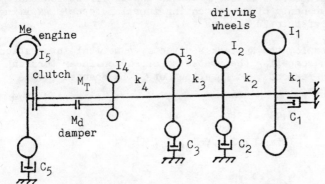

Figure 3. Equivalent Model for Torsional Vibration Analysis of the Truck NJD-131's Power Train.

In order to optimize the torsional damper's design project on the basis of dynamic response, the power train of the truck NJD-131 is simplified into an equivalently computational model for torsional vibration, as shown in Figure 3. The system's vibrational differential equations are:

$$\begin{cases} I_1\ddot{\varphi}_1 + C_1\dot{\varphi}_1 + k_1\varphi_1 - k_2(\varphi_2-\varphi_1) = 0 \\ I_2\ddot{\varphi}_2 + C_2\dot{\varphi}_2 + k_2(\varphi_2-\varphi_1) - k_3(\varphi_3-\varphi_2) = 0 \\ I_3\ddot{\varphi}_3 + C_3\dot{\varphi}_3 + k_3(\varphi_3-\varphi_2) - k_4(\varphi_4-\varphi_3) = 0 \\ I_4\ddot{\varphi}_4 - M_d + k_4(\varphi_4-\varphi_3) - M_T = 0 \\ I_5\ddot{\varphi}_5 + C_5\dot{\varphi}_5 + M_d + M_T = M_e \end{cases} \quad (22)$$

where

$$\begin{cases} M_d = M_\mu \cdot \text{sgn}(\dot{\beta}) = M_\mu \cdot \text{sgn}(\dot{\varphi}_5 - \dot{\varphi}_4) \\ M_T = \tilde{m}_t + \tilde{k}_5(\varphi_5 - \varphi_4) \quad (\varphi_5 - \varphi_4 = \beta) \end{cases} \quad (23)$$

$$M_e = \sum_{i=1}^{4} M_{ei} \quad (24)$$

$$M_{ei} = M_{gi} + M_{ji}$$

$$= M_{g0}\eta_m + \sum_{i=1}^{6} \eta_m M_{g,2\nu} \sin(2\nu\omega t + \delta_{g,2\nu} - 2\nu\theta_{1i})$$

$$- \frac{G_j}{g} R^2\omega^2\eta_m [\frac{\lambda}{4}\sin(\omega t - \theta_{1i}) - \frac{1}{2}\sin(2\omega t - 2\theta_{1i})$$

$$- \frac{3\lambda}{4}\sin(3\omega t - 3\theta_{1i}) - \frac{\lambda^2}{4}\sin(4\omega t - 4\theta_{1i})]$$

Equation 23 expresses the damper's damping and elastic characteristics; refer to Figures 1 and 2, and Lü Zhenhua[1]. Equation 24 includes the first six principal harmonic components of gaseous pressure's exciting torque and the first four harmonic components of reciprocating inertial force's exciting torque; refer to Lü Zhenhua[1].

We employ the Newmark's directly numerical integration method in computing the system's forced nonlinear torsional vibration response. So Equation Set 22 is changed to the matrix form:

$$I\{\ddot{\varphi}\} + C\{\dot{\varphi}\} + K\{\varphi\} = \{T\} \quad (25)$$

where $I = \text{Diag}[I_1, I_2, I_3, I_4, I_5]$

$C = \text{Diag}[C_1, C_2, C_3, 0, C_5]$

$$K = \begin{bmatrix} k_1+k_2 & -k_2 & & & \\ -k_2 & k_2+k_3 & -k_3 & & \\ & -k_3 & k_3+k_4 & -k_4 & \\ & & -k_4 & k_4+\tilde{k}_5 & -\tilde{k}_5 \\ & & & -\tilde{k}_5 & \tilde{k}_5 \end{bmatrix}$$

$$\{T\} = \begin{Bmatrix} 0 \\ 0 \\ 0 \\ M_d + \tilde{m}_t \\ M_e - M_d - \tilde{m}_t \end{Bmatrix}$$

In this example, the divisional objective functions f_d, f_v, and f_a in Equation 6 are taken to be the r.m.s. values of the first m maximum φ_i, $\dot{\varphi}_i$, and $\ddot{\varphi}_i$ respectively; i.e.,

$$\begin{cases} f_d(|\varphi_i(X,t)|_{max}) = \left[\frac{1}{m}\sum_{j=1}^{m}|\varphi_i|_{max,j}^2\right]^{\frac{1}{2}} \\ f_v(|\dot{\varphi}_i(X,t)|_{max}) = \left[\frac{1}{m}\sum_{j=1}^{m}|\dot{\varphi}_i|_{max,j}^2\right]^{\frac{1}{2}} \\ f_a(|\ddot{\varphi}_i(X,t)|_{max}) = \left[\frac{1}{m}\sum_{j=1}^{m}|\ddot{\varphi}_i|_{max,j}^2\right]^{\frac{1}{2}} \end{cases} \quad (26)$$

$$i=1, 2, \cdots, n \ (=5)$$

Now because of lacking the torsional vibration norm about automotive power train, the relative angle displacement constraints of Equation 7 are inconvenient to consider, so are omitted temporarily.

Also because of lacking enough design and application experiences on the nonlinear torsional damper at present, the constraints of Equation 9 are omitted as well, which only affects the CPU time spent on the optimization computation but does not influence the optimal solution.

On the other hand, the natural torsional vibration characteristic of the truck NJD-131's power train determined by means of a model with ten degrees - of - freedom shows the following points:

1) Changing the damper's torsional stiffness k_t may only adjust the system's third, and one of fifth, sixth, and seventh, natural frequencies;

2) In the torsional stiffness k_t's value range allowed by structural design, the system's third natural frequency is lower, generally f_{n3}= 20 to 40 Hz, and so the vehicle's principal critical speed interval corresponding to the resonances between the third natural vibration and the engine's second or fourth harmonically exciting torque component is brought down under the frequently used vehicle speed range, that is V_a< 20 km/h for the third gear and V_a< 30 km/h for the fourth or top gear.

3) The other resonant frequency that k_t may influence corresponds to the engine's harmonically exciting torque components the orders of which are not less than six (namely, $v \geqslant 6$), therefore really there is not any danger.

So, in the optimization design of the damper, the constraints about the system's natural frequencies may be omitted quite rationally, and the corresponding part of computational amount may be cut down.

It is clear that when the engine's speed and load, gearbox's gear, vehicle's weight, etc. are changed, the optimization results of the damper design are not the same. Hence the optimization should be carried on separately under several selected driving conditions (that can be called referential working conditions), and then an optimal project is determined by a comprehensive analysis. Taking the truck NJD-131 for an example, we may generally choose the referential driving conditions as: vehicle with the zero-loaded weight, engine under the frequently used load (generally (50 to 70)%) and speed range, and gearbox at the third or the fourth gear. In addition, we may also select the engine's speed at the system's principal critical speed points of the fourth resonance. Because the system's fourth natural frequency changes very little with the damper's torsional stiffness k_t, it can be computed by means of the system's equivalent torsional vibration model without the damper.

Yet for the optimization algorithm we have chosen we may find a locally optimal solution. In order to obtain the globally optimal solution, several optimization computations have to be carried out under different initial conditions, and eventually the truly optimal result is taken as the optimized design project of the damper.

Given for instance, a referential driving condition is selected as: vehicle with no loaded weight, engine under a 50% load and at 2400 rpm speed (which is the maximum torque point on the engine's full load characteristic curve), and gearbox at the fourth gear.

As has been stated above, some parameters are beforehand assigned values as follows:

$\beta_j = 8.0°$, $\beta_j' = 6.0°$, $\beta_{t1} = \beta_t' = 2.0°$,

$M_{n1} = M_{n1}' = 0.5$ kg·m, $M_{t1} = M_t' = 3.0$ kg·m,

$M_{\mu 1} = 0.5$ kg·m

The upper and lower bounds of design variables are given respectively as

$$X_l = \{M_{j1}, \beta_{t21}, M_{t21}, M_{n21}, M_{n31}, M_{\mu 21}\}^T$$

$$= \{20.0 \text{ kg·m}, 3.0°, 5.0 \text{ kg·m}, 0, 0, 0\}^T$$

$$X_u = \{M_{ju}, \beta_{t2u}, M_{t2u}, M_{n2u}, M_{n3u}, M_{\mu 2u}\}^T$$

$$= \{40.0 \text{ kg·m}, 7.0°, 30.0 \text{ kg·m}, 5.0 \text{ kg·m},$$

$$5.0 \text{ kg·m}, 5.0 \text{ kg·m}\}^T$$

According to our experience, the initial values of design variables are given as

$$X^{(0)} = \{30.0 \text{ kg·m}, 5.0°, 15.0 \text{ kg·m}, 1.0 \text{ kg·m},$$

$$1.0 \text{ kg·m}, 1.0 \text{ kg·m}\}^T$$

In Equation Set 26, we make m = 10. And in the objective function Equation 6, all of the weight factors' values are assigned as

$$W_d = 10.0, \quad W_v = 0.0, \quad W_a = 10^{-4},$$

i	1	2	3	4	5
W_{di}	1.0	1.0	1.0	0.1	0.0
W_{ai}	100.0	10.0	10.0	1.0	0.0

The penalty factor's initial value and its reducing coefficient are given as

$$r^{(0)} = 2.5, \quad C = 0.2.$$

We have realized the above optimization computation procedure with a Dual 68000 computer in FORTRAN 77 programming language, and the programme is named OPTVDC, in which the simulant computation programme FNLTVS of forced nonlinear torsional vibration of automotive power train was used. For the latter programme we shall give a special study in another paper.

The solution of the above example problem is

$$X^* = \{M_j^*, \beta_{t2}^*, M_{t2}^*, M_{n2}^*, M_{n3}^*, M_{\mu 2}^*\}^T$$

$$= \{33.5 \text{ kg·m}, 5.4°, 17.2 \text{ kg·m}, 0.2 \text{ kg·m},$$

$$0.4 \text{ kg·m}, 1.2 \text{ kg·m}\}^T$$

OUTLOOK

As has been stated in the opening section of this paper, because the first stage of a triple stage damper has different functions from the second and the third stages, different mathematical models may be constructed in order to optimize their performance parameters. So far as the second and the third stages of the damper are concerned, they have the functions of not only damping torsional vibration when the automobile is properly travelling but also absorbing torsional shock when automobile is starting, shifting gears, or braking. The latter is also an important aspect to be considered in the optimization design of the damper.

The optimization design of dynamic performance parameters of the damper is certainly important, but only when developed to the optimization of structural parameters, can it act directly on productive practice.

Because of the complexity and difficulty of the optimization problem based on dynamic response of nonlinear vibrational system, and also because any algorithm we may employ in solving a constrained nonlinear programming problem can not be universally effective, it is not easy to select a proper algorithm and to realize effectively and economically the comprehensive optimization of the nonlinear torsional multiple stage damper.

To sum up, we still have much investigatory work to do for optimization of the damper design.

REFERENCES

1. Lü Zhenhua (1985), A Study of Nonlinear Torsional Multiple Stage Damper in Clutch of Truck NJD-131, Thesis for the Master degree, Jilin University of Technology.

2. Xi Shaolin and Zhao Fengzhi (1983), Computational Methods for Optimization, Shanghai Science & Technology Publication.

3. Wayne V. Nack (1984), Optimization for Vibration Isolation, International Journal for Numerical Methods in Engineering, Vol.20.

4. Chen Lizhou, Zhang Yinghui, and others (1982), Mechanical Optimum Design, Shanghai Science & Technology Publication.

5. C.C. Hsieh and J.S. Arora (1984), Design Sensitivity Analysis and Optimization of Dynamic Response, Computer Methods in Applied Mechanics and Engineering, Vol.43.

SECTION 3 MANUFACTURE

A Numerical Tool for CAD of Sheet Metal Forming Process in Automotive Industry
B.İ. Kılkış
Department of Mechanical Engineering, Middle East Technical University, Ankara, Turkey

INTRODUCTION

Body building in Automotive industry involves different formed sheet metal parts in a typical vehicle. Although the outer shape gets more and more aerodynamic and thus sharp corners and large bends are eliminated in recent models, and some parts are being replaced by plastic moldings, sheet metal forming is still a challenging problem for engineers in the sense that better tooling and efficient material expenditure is the primary goal in the process.

New plastic flow equations in cartesian co-ordinates were developed originally for the radial-drawing region of non-circular deep-drawn cups for 2-D stress condition by Kaftanoğlu and Kılkış[1,2]. Using this model, the stresses, strains and therefore the absolute displacements of material points can be determined in this region using finite difference and inverse interpolation techniques. The radial region is still important for the laymen in order to predict the final rim geometry or vice versa. This is important in minimizing the amount of scrap both due for quality control returns and extra rim to be chopped after stamping.

The computer model discretizes the radial region of the blank to be stamped or deep-drawn such that both time and material non-linearities are physically followed throughout the stamping or drawing process. At each increment of the punch load, computer output shows the stress strain values at each point. This information is relevant to predict early material failures and to update the outer rim geometry. The solution has to be started from a line of symmetry with respect to stress. Therefore even though the geometry should not necessarily be circular, in the current algorithm, a stress symmetry line should be present in the specific part to be analysed.

The computer program can be readily applied to stamped or deep-drawn parts in automotive industry. The program inputs are the material properties, strain hardening characteristics of the material, symmetry conditions & initial boundary conditions for stress at the flange inner edge. The solution algorithm, being virtually an iterative one, takes a long time which makes it difficult for an actual real time analysis. However, on line terminals can be used for data entry, geometrical inputs and graphical display of the final flange geometry and the rim contour. A simple technique for the laymen has also been developed by Kılkış[1] in predicting the final rim geometry using slip-line techniques.

The paper includes the basic theory, stress-strain and equilibrium equations for cartesian co-ordinates and describes the basic algorithm of the slip-line solution with a case study.

DEVELOPMENT OF THE THEORY

A theory is developed by Kaftanoğlu and Kılkış[3] for plane stress problems in the plastic deformation regime. Assumptions employed are:
- Thickness direction is a principal direction,
- Stresses vary linearly in infinitesimally small time increments,
- Thickness is variable.

Plastic strain definitions
Plastic strains are defined along initially orthogonal lines namely α and β. After an infinitesimal deformation, these lines are expected to deform into curvilinear and non-orthogonal lines. Strains along these lines and relevant nomenclature is given in Figure 1. Due to small material thickness, the analysis of the non-orthogonality is confined to the α-β plane.

In this Figure, broken lines show an arbitrarily deforming small grid element A'B'C'D' at n+1 th. stage of deformation. Grid element ABCD is the original grid element prior to deformation. If the strain rate of any point such as D' is of interest, all the orthogonal components of incremental strains should be known at any given instant. Superscript (') is associated with expressions related with the non-orthogonal element. Let the deformation at time step (n+1) at point D' be known and point D and D' be made to coincide in order to analyse the strains of this specific infinitesimal element. Subtended angle ξ_4 at this point D:

$$\xi_4 = 90° - \psi_\beta + \psi_\alpha' \tag{1}$$

Shear strains Partial incremental plastic shear strains associated with the α' and β-lines at time n+1 are defined

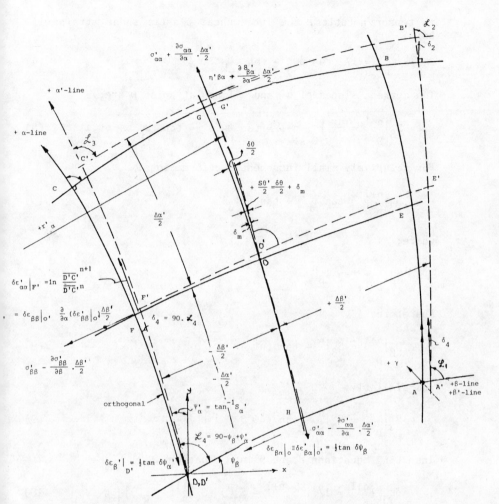

Figure 1. An arbitrarily deformed, non-orthogonal grid element.

accordingly as:

$$\delta\gamma_\beta|^{n+1} = \tan\psi_\beta^{n+1} \qquad (2\text{-}a)$$

$$\delta\gamma_\alpha|^{n+1} = \tan\psi_\alpha^{n+1} \qquad (2\text{-}b)$$

Here ψ_β is measured from the positive x axis and ψ_α is measured from the positive y axis to the current tangent of α' and β-line respectively. The total incremental plastic shear strain at point D is:

$$(\delta\gamma_{\beta\alpha}|^{n+1})_D = (\tan\delta\psi_\beta^{n+1} - \tan\delta\psi_\alpha^{n+1}) \qquad (3)$$

in tensor notation, the incremental plastic shear strain will be:

$$\delta\varepsilon_{jk} = 1/2(\tan \delta\psi_j - \tan \delta\psi_k) \qquad (4)$$

The current slopes of α' and β-lines at point D are:

$$S_\beta\Big|_D^{n+1} = \frac{dy}{dx}\Big|_{D-A}^{n+1} = \tan \psi_\beta^{n+1} = \tan(\psi_\beta^n + \delta\psi_\beta) \qquad (5)$$

For relatively small incremental deformations:

$$S_\beta\Big|_D^{n+1} \cong \tan \psi_\beta^n + \tan(\delta\psi_\beta) \qquad (6)$$

where;

$$S_\beta\Big|_D^n = \tan \psi_\beta^n \qquad (7)$$

Therefore;

$$S_\beta\Big|_D^{n+1} \cong S_\beta\Big|_D^n + \tan(\delta\psi_\beta) \qquad (8)$$

and similarly:

$$S_\alpha\Big|_D^{n+1} \cong S_\alpha\Big|_D^n + \tan(\delta\psi_\alpha) \qquad (9)$$

Inserting Equations 8 and 9 into Equation 4:

$$\delta\varepsilon_{jk}\Big|_D^{n+1} \cong 1/2\big|(S_j^{n+1} - S_j^n) - (S_k^{n+1} - S_k^n)\big|_{\substack{j\equiv\beta\\k\equiv\alpha}} \qquad (10)$$

The last equation simply states that any incremental shear strain at a point taking place between successive and small deformation stages is the difference of slope changes in the respective sides. It is evident from Figure 1 that α'-lines are close to orthogonal direction and consequently ψ_α angles are close to 90 degrees. In numerical computations; this may cause certain troubles dealing with such angles, therefore it is more convenient to write Equation (10) in the following manner:

$$\delta\varepsilon_{\beta\alpha}\Big|_D^{n+1} \cong 1/2\big|(S_\beta^{n+1} - S_\beta^n) + (S_\alpha^{*n+1} - S_\alpha^{*n})\big| \qquad (11)$$

where

$$S_\alpha^*\Big|_D^{n+1} = \frac{dx}{dy}\Big|_{D-C^{n+1}} \tag{12}$$

and ψ_α^* angles are measured from the positive y-axis.

Normal strains For infinitesimally small lengths (1), the incremental normal plastic strain is given by the logarithmic definition:

$$\delta\varepsilon_{ii}^{n+1} = \ln \frac{l^{n+1}}{l^n} \tag{13}$$

The last equation describes the strains in a linear element of small length. For curvilinear lines, the definition becomes more complicated. Although the arc can be approximated by infinitesimally small straight lines, their direction will vary. Summing up the individual, infinitesimal, "linear", plastic strains, only an "average" normal strain associated with the given arc length can be obtained. Then:

$$\delta\tilde{\varepsilon}_{\beta\beta}\Big|_{D-A^{n+1}}^{n+1} = \delta\varepsilon_{\beta\beta}\Big|_H^{n+1} = \ln \frac{\overline{D-A}^{n+1}}{\overline{D-A}^n} \tag{14}$$

This equation states that the average incremental plastic strain $\delta\tilde{\varepsilon}_{\beta\beta}$ along the arc length is numerically equal to the infinitesimal plastic strain $\delta\varepsilon_{\beta\beta}$ at the mid-point of the arc length. Similarly:

$$\delta\tilde{\varepsilon}'_{\alpha\alpha}\Big|_{D-C^{n+1}}^{n+1} = \delta\varepsilon'_{\alpha\alpha}\Big|_F^{n+1} = \ln \frac{\overline{D-C}^{n+1}}{\overline{D-C}^n} \tag{15}$$

Allowing a variation for the normal strains along the half lengths of arcs D-C and DA:

$$\delta\tilde{\varepsilon}_{\beta\beta}\Big|_D^{n+1} \cong \delta\varepsilon_{\beta\beta}\Big|_H^{n+1} - \frac{\partial}{\partial\beta}(\delta\varepsilon_{\beta\beta}\Big|_H^{n+1}) \cdot \frac{\Delta\beta}{2} \tag{16}$$

$$\delta\varepsilon'_{\alpha\alpha}\Big|_D^{n+1} \cong \delta\varepsilon'_{\alpha\alpha}\Big|_F^{n+1} - \frac{\partial}{\partial\alpha}(\delta\varepsilon'_{\alpha\alpha}\Big|_F^{n+1}) \cdot \frac{\Delta\alpha'}{2} \tag{17}$$

Different forms of variations could be proposed, but a linear variation is found out to be reasonably accurate and time saving for numerical calculations. From constancy of volume one can determine the normal strain in thickness direction:

$$\delta\varepsilon_{\alpha\alpha} + \delta\varepsilon_{\beta\beta} + \delta\varepsilon_{\gamma\gamma} = 0 \tag{18}$$

Therefore;

$$\delta\varepsilon_{\gamma\gamma}\Big|_D^{n+1} = -(\delta\varepsilon_{\alpha\alpha}\Big|_D^{n+1} + \delta\varepsilon_{\beta\beta}\Big|_D^{n+1}) \tag{19}$$

where $\delta\varepsilon_{\alpha\alpha}\Big|_D^{n+1}$ is the orthogonal component of $\delta\varepsilon'_{\alpha\alpha}\Big|_D^{n+1}$.

Apart from the oblique components of incremental strains, their equivalent orthogonal components can be easily computed by strain Mohr circle using the known or assumed subtended angle ξ at point D, whenever necessary for the utilization of the orthogonal plasticity equations related to a material point in the deformation field. This is shown in Figure 2.

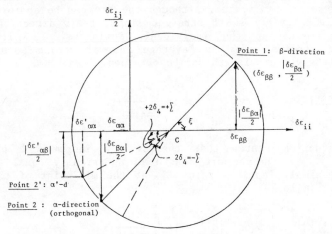

Figure 2. Mohr circle representation of the strain state.

Equations 11, 16, 17 and 19 give the state of strain at any point in the deformed field provided that the bounding arcs are known or assumed as shown in Figure 4.

Basic Equations
Apart from the constancy of volume as given in Equation 18, following equations are valid for a material point in an orthogonal system of co-ordinates:

Yield criterion by Von Mises [4]:

$$(\sigma_{\alpha\alpha}-\sigma_{\beta\beta})^2+(\sigma_{\beta\beta}-\sigma_{\gamma\gamma})^2+(\sigma_{\gamma\gamma}-\sigma_{\alpha\alpha})^2+6(\tau_{\alpha\beta}^2+\tau_{\beta\gamma}^2+\tau_{\gamma\alpha}^2)=\bar{\sigma}^2 \tag{20}$$

Here σ stands for normal stress components defined in an orthogonal system of co-ordinates with α, β and γ axes. Shear stresses are denoted by τ, and $\bar{\sigma}$ is the equivalent stress.

From Swift's law for a strain-hardening material:

$$\bar{\sigma} = A(B+\bar{\varepsilon})^n \quad (A,B,n = \text{material properties for hardening}). \tag{21}$$

The equivalent strain $\bar{\varepsilon}$ is the integral sum of incremental equivalent strains calculated over the actual strain path:

$$\bar{\varepsilon} = \int_0^{\bar{\varepsilon}} d\bar{\varepsilon} \quad \text{and;} \tag{22}$$

$$d\bar{\varepsilon} = \sqrt{2}/3 \{(\delta\varepsilon_{\alpha\alpha}-\delta\varepsilon_{\beta\beta})^2 + (\delta\varepsilon_{\beta\beta}-\delta\varepsilon_{\gamma\gamma})^2 + (\delta\varepsilon_{\gamma\gamma}-\delta\varepsilon_{\alpha\alpha})^2 + 6(\delta\varepsilon_{\alpha\beta}^2 + \delta\varepsilon_{\beta\gamma}^2 + \delta\varepsilon_{\gamma\alpha}^2)\}^{1/2} \tag{23}$$

Plastic flow rule [4]:

$$\frac{\delta\varepsilon_{\alpha\alpha}-\delta\varepsilon_{\beta\beta}}{\sigma_{\alpha\alpha}-\sigma_{\beta\beta}} = \frac{\delta\varepsilon_{\beta\beta}-\delta\varepsilon_{\gamma\gamma}}{\sigma_{\beta\beta}-\sigma_{\gamma\gamma}} = \frac{\delta\varepsilon_{\gamma\gamma}-\delta\varepsilon_{\alpha\alpha}}{\sigma_{\gamma\gamma}-\sigma_{\alpha\alpha}} = \frac{\delta\varepsilon_{\alpha\beta}}{\tau_{\alpha\beta}} = \frac{\delta\varepsilon_{\beta\gamma}}{\tau_{\beta\gamma}}$$

$$= \frac{\delta\varepsilon_{\gamma\alpha}}{\tau_{\gamma\alpha}} = \delta\lambda, \tag{24}$$

Using the principle of constancy of volume, and making the following substitutions, $\delta\lambda$ term which is a variable parameter can be eliminated:

i- $X = \sigma_{\alpha\alpha}/\sigma_{\beta\beta}$; $Z = \tau_{\alpha\beta}/\sigma_{\alpha\alpha}$; $\sigma_{\gamma\gamma} = 0$ (plane-stress) (25)

ii- $\tau_{\beta\gamma} = \tau_{\gamma\alpha} = 0$ (for thin sheet of materials) (26)

Consequently;

$$X = \frac{2.\delta\varepsilon_{\alpha\alpha}+\delta\varepsilon_{\beta\beta}}{2.\delta\varepsilon_{\beta\beta}+\delta\varepsilon_{\alpha\alpha}} \quad \text{and} \quad Z = \frac{\delta\varepsilon_{\alpha\beta}}{2.\delta\varepsilon_{\alpha\alpha}+\delta\varepsilon_{\beta\beta}} \tag{27}$$

$$\delta\bar{\varepsilon} = \frac{2.\delta\varepsilon_{\alpha\alpha}}{(2X-1)} \cdot \{X^2-X+1\}^{1/2} \tag{28}$$

$$\sigma_{\alpha\alpha}\{1-1/X+1/X^2+3Z^2\}^{1/2} = A.(B+\bar{\varepsilon})^n \tag{29}$$

In conclusion, the normal components of stresses at a point can simply be expressed in terms of associated oblique plastic strains for a yielding material provided that the history of deformations are followed.

Field equations of equilibrium:

Although the classical equations of plasticity can still be used for point variables for a non-orthogonal state, provided that necessary transformations are made; the actual deformation pattern should be considered in deriving the field equations of equilibrium. The following stress equations of equilibrium were derived for such a state under plane stress conditions:

$$\frac{\partial}{\partial \beta}(\sigma_{\beta\beta} \cdot t) + \frac{\partial}{\partial \alpha}(\tau'_{\beta\alpha} \cdot t) + \frac{t \cdot (\sigma_{\beta\beta} - \sigma'_{\alpha\alpha})}{r'_\alpha} - \frac{t}{r_\beta} \cdot (\tau'_{\beta\alpha} + \tau'_{\alpha\beta}) = 0$$
$$\beta\text{-direction} \quad (30)$$

$$\frac{\partial}{\partial \alpha}(\sigma'_{\alpha\alpha} \cdot t) + \frac{\partial}{\partial \beta}(\tau'_{\alpha\beta} \cdot t) + \frac{t(\sigma'_{\alpha\alpha} - \sigma_{\beta\beta})}{r_\beta} + \frac{t}{r'_\alpha} \cdot (\tau'_{\alpha\beta} + \tau'_{\beta\alpha}) = 0$$
$$\alpha'\text{-direction} \quad (31)$$

Here t is the thickness and is allowed to change linearly along the grid element section. It should be noted that $\tau'_{\alpha\beta}$ is not equal to $\tau'_{\beta\alpha}$. With the substitutions: $U' = \sigma'_{\alpha\alpha} \cdot t$; $V = \sigma_{\beta\beta} \cdot t$; $T = \tau'_{\beta\alpha} \cdot t$; $T' = \tau'_{\alpha\beta} \cdot t$:

$$\frac{\partial}{\partial \beta}(V) + \frac{\partial}{\partial \alpha}(T) + \frac{V \cdot (1-X)}{X \cdot r'_\alpha} - \frac{1}{r_\beta} \cdot (T+T') = 0 \quad \beta\text{-d} \quad (32)$$

$$\frac{\partial}{\partial \alpha}(U') + \frac{\partial}{\partial \beta}(T') + \frac{U' \cdot (1-X)}{r_\beta} + \frac{1}{r'_\alpha} \cdot (T'+T) = 0 \quad \alpha'\text{-d} \quad (33)$$

The last two equations, together with the other relevant equations can be effected to compute the actual stresses at the next grid-point. The orthogonal components of stresses can be computed by using ξ angle and the stress Mohr-circle whenever necessary such as for the yield Equation given in Equation 20. Figure 3 shows the stress Mohr circle.

Although the above mentioned equations are sufficient for a step by step solution, it is not so easy to develop an effective numerical algorithm due to non-linear nature of the problem. Figure 4 shows a general non-orthogonal grid element where the deformation variables at sides D'-C', and A'-D' are either known (as boundary conditions) or fixed (previously solved points).

Step-by step solution of deformation space at time n

Starting from an initially undeformed blank, the plane stress problem is analysed at discrete intermediate stages. Points D', A', and C' are known from previous solution or boundary conditions, and the problem is to compute the new position of B (B') at the given stage[2]:

CAD OF SHEET METAL FORMING PROCESS 97

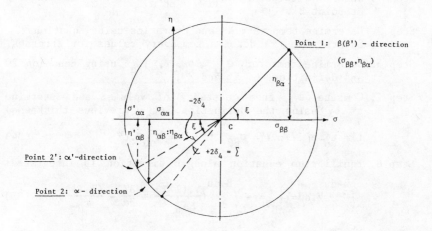

Figure 3. Stress Mohr circle.

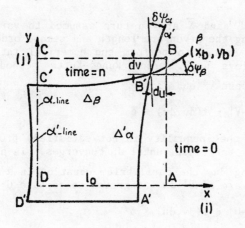

Figure 4. Deformation of the initial grid.

In this figure, du and dv are the trial displacement values of point B.

Step 1. Assume dv (outer loop)

Step 2. Assume du (inner loop)

Step 3. Fit second-order polynomials to arcs C'-B' and A'-B' using assumed displacements du and dv.

Step 4. Determine temporary $\delta\varepsilon'_{\alpha\alpha}$, $\delta\varepsilon_{\beta\beta}$, $\delta\varepsilon'_{\beta\alpha}$, $\delta\varepsilon'_{\alpha\beta}$ and angle ξ at point B'.

Step 5. Determine temporary X* and Z* ratios using Equation 27. Superscript (*) denotes temporary values for stresses.

Step 6. Determine temporary $\sigma'_{\alpha\alpha}*$, $\sigma_{\beta\beta}*$, $\tau_{\beta\alpha}*$ using Equation 29 and definitions of X* and Z*.

Step 7. Compute $\delta\varepsilon_{\gamma\gamma}$ from constancy of volume, and determine t* by using the definition: (t_o = previous thickness)

$$t^* = t_o \cdot e^{\delta\varepsilon_{\gamma\gamma}}; \quad V^* = \sigma_{\beta\beta}^* \cdot t^*; \quad U^* = \sigma'_{\alpha\alpha}^* \cdot t^* \quad (34)$$

Step 8. Equilibrium equation along β-direction is integrated:

$$\int_{C'}^{B-du} V \cdot d\beta = V\Big|_{B'} = -\int_{C'}^{B-du} \left(\frac{V \cdot (1-\overline{X})}{r'_\alpha} + \frac{\partial T}{\partial \alpha} - \frac{(T+T')}{r_\beta}\right) d\beta \quad (35)$$

where

$$\overline{X} = \frac{X^* + X^o}{2} \quad (36)$$

T, T' and $\overline{X}$ are in turn assumed to vary non-linearly along the given arc length and second order polynomials are fitted to perform the numerical integration which employs 4.th. order Runge-Kutta method to solve V.

Step 9. Compare V* and V:

$$|V^* - V| = f(du, dv) < \delta \quad (37)$$

If the inequality is not satisfied, procedure returns back to step.2. until du converges. δ is a small number.

Step 10. Using du, the equilibrium equation in α'-direction for the arc A'-B' is integrated to compute U:

$$|U^* - U| = f(dv, du) < \delta \quad (38)$$

Procedure returns back to step.1., if the above inequality is not satisfied.

Step 11. Converged values for du, dv are thus found and the position of C is fixed.

Step 12. Converged stress and strain values along the arcs A'-B' and C'-B' are fixed and the program shifts to the next grid point.

SAMPLE SOLUTION

The flange region of a deep-drawn square blank with 0.99 mm. thickness and 80x80 mm. dimensions is analysed as a case study. The square punch has the dimensions of 40x40 mm. Due to the diagonal symmetry, only the 1/8 th. of the flange region is solved. Solution starts from the die inlet at point 2. The next point along the axial-symmetry line 2-1 is first solved, then the grid points on a single β-line is solved until the diagonal is reached. The procedure repeats itself until the rim is reached.

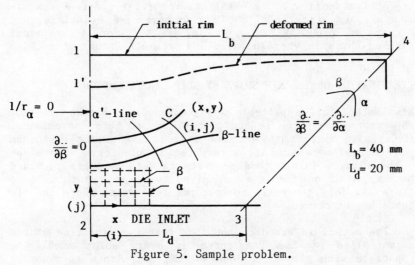

Figure 5. Sample problem.

Figure 6 gives a sample solution where the rim shape is predicted at different stages (NS). The results were compared by experiments being performed on a Tinius-Olsen sheet metal testing apparatus. The computer program oriented for Computer

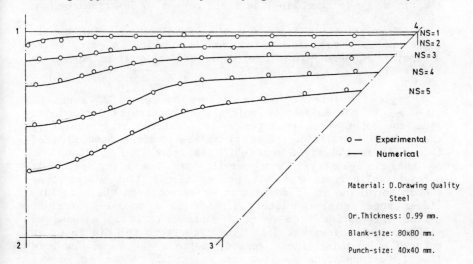

Figure 6. Determination of the shape of the rim.

100 CAD OF SHEET METAL FORMING PROCESS

Aided Design consists of the main solution program and a graphics subroutine to display the displacement solutions of the radial region of a deep-drawn or stamped sheet metal. As the solution algorithm is an iterative one for each material point at a given time step, true interaction is not possible due to time delay. However on fast computers, an acceptable interaction speed will be obtained. For first order approximations a slip line technique is also developed by Kılkış[1] and a computer program has been implemented for a graphical solution which is very suitable and accurate enough at machine shop level.

DETERMINATION OF THE RIM SHAPE BY SLIP-LINE TECHNIQUE

This method is applicable to rectangular blanks of any proportion as well as square blanks. Due to this convenience, it is believed that this method will be useful for engineers and die designers who are interested in predicting the shape of the flange. This method predicts the rim geometry provided that a point on the rim is given such as point A. (See Figure 7). This generally corresponds to a point where the rim starts to get curved.

The method is based on the fact that the direction of the tangent line of the rim curve at every point should be compatible with the directions of the slip-lines terminating on the rim. Rim is a free boundary and in the limit, it can be visualized as a principal stress direction at every point. As a consequence, any slip-line should terminate on the rim such that it will make 45 degrees with the rim tangent at this point. This necessary condition is also sufficient to construct the rim shape in the slip-line field region as described below:

A new slip-line field for a D.R. value less than two is suggested so as to conform to the rim shape which is going to be determined. There are three distinct regions as shown in Figure 7:

Region 1: This region is defined by points 3, 1 and C. Line 3-1 is the well-known velocity discontinuity line. Point A is the point which is going to be predicted analytically by the slip-line solution given by Kılkış[1]. Right hand side of this line is the rigid zone which is prescribed by line X-X. As the rim geometry is a curvilinear line after point A, straight slip-lines can not extend beyond line AC as they cannot terminate at 45 degrees at every point on the rim. This linear zone is appropriately limited by a bounding straight β-line originating from point A where the rim is assumed to start to curve inwards. This line intersects the die inlet at point C. In this linear region, α-lines are shown in bold lines. Corresponding β-lines are partially shown in broken lines at the die inlet.

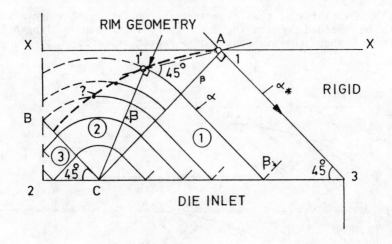

Figure 7. Slip-line field for a curved rim shape
(Reduction Ratio (R.R.) < 1.5).

Region 2: This region consists of circular fans with their centers at point C. They continue down to the straight line C-B which separates regions 2 and 3. Circle arcs correspond to α-lines and their radii correspond to β-lines.

Region 3: This region is again a linear region. Circular α-lines coming from region 2 continue in their tangential direction as straight lines until they reach up to die inlet or axial-symmetry line.

Broken lines show the possible geometry of the rim in region 2. The sufficient condition to determine the geometry is formulated as follows: If this region is sub-divided into sufficiently small increments, then the rim curve can be approximated by straight lines. Such a straight line is shown for an exaggerated incremental region in Figure 7: Line 1-1' representing the rim in this region should be so oriented that the angle between a circular α-line passing through point 1' should be 45 degrees. This uniquely fixes the position of point 1' on the rim. Other successive points can be determined with the same reasoning. After all the points are determined, a smooth curve is passed through them. This curve will define the rim shape. Figure 8 shows typical solutions for R.R. values of 1.4 and 1.80 respectively. This figure shows that the rim can not reach the axial-symmetry line 2-1 for R.R.= 1.4. Figure 8 shows the experimental points also for the same specimen for drawing ratios of 1.42 and 1.80. Here, it may be noted that the center point C shifts to the axial-symmetry line for R.R. values greater than 1.5. In this case region 3 vanishes. In the special case when R.R. is equal to 1.5, point C coincides with point 2. When the center point shifts to the axial-symmetry line, circular α-lines can not conform to the boundary condition

102 CAD OF SHEET METAL FORMING PROCESS

in the vicinity of this boundary. The condition of singularity can be avoided by leaving an infinitesimal gap of length δ. Figure 9 shows the rim shape for a reduction ratio of 1.7.

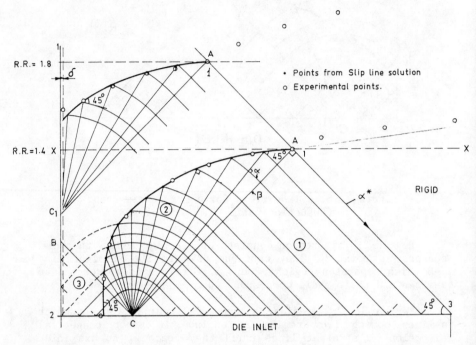

Figure 8. Determination of rim shape. R.R. 1.40 and 1.80.

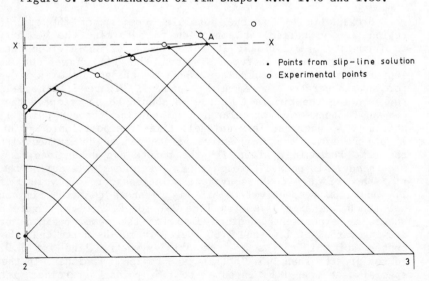

Figure 9. Rim shape for R.R. = 1.7.

CONCLUSIONS

Two techniques are presented for the deformation analysis in the flange region of sheet metal forming process. The numerical solution has been tested for different cases. Satisfactory results have been obtained. Due to its iterative nature this technique is more suitable for long term research and development type achivities in the automotive industry.

The second method which is a graphical technique is fast, simple and sufficiently accurate for everyday use at machine shop level.

REFERENCES

1. Kılkış, B. (1980), Theoretical and Experimental Investigation of the Radial-Drawing Region of Square Blanks, Ph.D. Thesis, METU, Ankara.

2. Kaftanoğlu, B. and Kılkış, B. (1980), Theory of Deep-Drawing of Square Blanks and the Numerical Solution of the Flange Region, in Memories Scientiefigues Revue Metallurgie, pp. 126 to 135, Proceedings of the 11th IDDRG Congress, Metz, France.

3. Kılkış, B. and Kaftanoğlu, B. (1980), On the Finite Difference Formulation of Non-linear Plastic Deformations in Deep-Drawing, Proceedings of the International Conference on Numerical Methods for Non-linear Problems, (Ed. K. Morgan, R.W. Lewis), pp. 210 to 221, Swansea, United Kingdom. Pineridge Press, Swansea.

4. Ford, H. (1963), Advanced Mechanics of Materials, Longmans, London.

Computational Verification of Numerical Control Programs for Sculptured Surface Parts
J.H. Oliver and E.D. Goodman
A.H. Case Centre for CAE/CAM, Michigan State University, East Lansing, Michigan 48824-1226, U.S.A.

INTRODUCTION

One of the most active research areas in computer-aided manufacturing deals with increasing the automation of numerically controlled (N/C) milling processes. This paper addresses the problem of verifying the correctness of N/C milling programs prior to actual milling. The problem of N/C verification is here dissected into two component parts: N/C simulation, which involves calculation of an "as milled" part model, and N/C geometric verification which involves comparison of an N/C program with an "as desired" model.

Since the dimensional quality of the N/C program is of primary concern, this work focuses on the latter of these component problems: specifically, a new definition of the N/C geometric verification problem. An algorithm which implements a solution is described briefly. Some results are presented which demonstrate solution of the verification problem in $O(N)$ time, (where N is the number of tool path points) which compares with $O(N^4)$ time complexity for previously presented techniques.

PROBLEM DEFINITION

Many high level general-purpose part-programming systems operate in two distinct steps; processing and postprocessing (Smith and Evans[1]). In the processing phase, mathematical models of the desired part, the milling tool, and generally, several limiting surfaces are used to generate an intermediate set of data points called the cutter location data, or CL-data. The CL-data file is sufficient to define the final shape (or geometry) of the desired part. Postprocessing involves incorporation of CL-data with machine specific factors such as tool feed and speed rates and coolant flow requirements. With some combinations of languages and milling machines, postprocessing may also involve the introduction of approximations needed to drive a particular mill: straight-line approximations to arcs, for instance.

Postprocessing parameters can, of course, affect the dimensional quality of the final part. For example, tool size and shape combined with feed and speed rates determine the material removal rate. The rate of material removal is limited by tool deflection, tool wear, and coolant flow rates. Postprocessing parameters are often determined based on conservative empirical data (tables) and the experience of the mill operator; they are typically verified via trial and error. Although these parameters are certainly an important part of the N/C program, their importance is secondary to the dimensional (or geometric) quality of the tool paths (the CL-data).

Methods previously proposed for automation of N/C verification (Voelcker and Hunt[3,4], Ruberl[5] and Fridshal[6]) rely on the direct application of solid modeling technology. These techniques approach the problem as a simulation problem, conducting all of the operations needed to check material removal rates, while interested only or primarily in the final geometric quality of the part. This leads to unnecessary constraints and a heavy computational load. In this work, the problem is broken into two parts: N/C geometric verification, in which the geometric quality of the milling program is checked, and N/C simulation, in which material removal rates and other such factors may be considered, but final geometric dimensions are not explicitly checked. (The term N/C simulation as used here implies calculation of either the union of all material removed or the workpiece remaining after each tool motion.) Dissection of the verification problem along these lines allows for solution of the N/C geometric verification problem independently of the N/C simulation problem. This paper deals exclusively with the N/C geometric verification problem.

<u>N/C geometric verification problem</u>
The following discussion provides a definition of the N/C geometric verification problem. Let,

> P represent a geometric description of trimmed, oriented surfaces comprising pertinent exterior portions of the desired milled part,
> H represent a geometric description of trimmed, oriented surfaces comprising pertinent exterior portions of the necessary holding fixtures, (which are not to be cut),
> N represent an N/C program including n distinct positions of the tool center, combined with appropriate specification of the paths to be followed between positions (i.e. interpolation rules),
> Q represent a tool geometry,

T_{in} be a uniform tolerance limit inside the desired part and,
T_{out} be a uniform tolerance limit outside the part.

Define the workpiece model as the union of the desired part and holding fixture models. For every point p_i on $\{P \cup H\}$, let ν_i represent the corresponding outward-directed surface unit normal vector.

For the jth step in program N, combine Q with N_j, N_{j+1} and the interpolation rule, to define S_j, the surface definition of the jth "swept volume".

Let $I_{i,j}$ equal the directed distance along ν_i from p_i to intersection with S_j. Then, define cut value,
$C(p_i) = \min_j \{ I_{i,j} \}$.

Define a <u>verification</u> mapping $V : P \to \{-1, 0, 1\}$ such that,

$$V(p_i) = \begin{cases} -1 \text{ if } C(p_i) < -T_{in} \\ 0 \text{ if } -T_{in} \leq C(p_i) \leq T_{out} \\ 1 \text{ if } T_{out} < C(p_i) \end{cases}$$

Also, define a <u>fixture cut</u> mapping $E : H \to \{-1, 0\}$ such that,

$$E(p_i) = \begin{cases} -1 \text{ if } C(p_i) < 0. \\ 0 \text{ otherwise.} \end{cases}$$

Then, the program N is <u>geometrically verified</u> iff for all $p_i \in P$, $V(p_i) = 0$, and for all $p_i \in H$, $E(p_i) = 0$. Program N <u>gouges</u> P if there exists p_i such that $V(p_i) = -1$, and <u>misses</u> P if there exists p_i such that $V(p_i) = 1$. The <u>gouged region</u> is defined as the set of points $R_g = V^{-1}(-1)$. Similarly, the <u>missed region</u> is defined as the set of points $R_m = V^{-1}(1)$.

Program N interferes with (cuts) holding fixtures H if there exists $p_i \in H$ such that $E(p_i) = -1$, and the <u>fixture collision region</u> is defined as the set of points $R_f = E^{-1}(-1)$.

This definition allows each point (theoretically infinitely many) on the workpiece model, p_i, to be affected by multiple tool motions (swept volumes, S_j), resulting in multiple intersections of the normal vector and swept volumes, $I_{i,j}$. However, only the closest or deepest excursion of the tool toward or into the surface is retained for each point as the cut value, $C(p_i)$. The cut values may then be evaluated to determine regions of the part model which have been gouged, missed, or cut within tolerance limits.

The tolerance model used in this definition is limited to surface features, i.e. T_{in} and T_{out} define a general unequally disposed bilateral tolerance zone relative to the desired part surface model. This tolerancing scheme conforms to the ANSI[6] standard Y14.5M-1982. However, characteristic tolerances relative to other geometric features (e.g., cylindricity, perpendicularity, concentricity, etc.) as well as material condition modifiers, are not addressed by this definition.

An interesting comparison can be drawn between this definition of the N/C geometric verification problem and another definition based on the direct application of solid modeling technology. Voelcker and Hunt[2,3] define the problem as a "null-object" calculation which is required when comparing an "as milled" solid model of the part with an "as desired" solid model. For example, let A represent the "as milled" part model and B represent the "as desired" part; then if the Boolean difference (A-B) is non-null, an undercut (missed) condition exists. Similarly, if the result of the difference (B-A) is non-null, an overcut (gouged) condition exists. The N/C program mills the desired part exactly if the symmetric difference, (A-B) U (B-A), is null. In the terminology of this dissertation, calculation of the "as milled" part is called N/C simulation, while the symmetric difference null-object calculation is called N/C geometric verification.

Although both definitions of the N/C geometric verification problem are useful, they are fundamentally different. For example, the definition presented by Voelcker and Hunt does not explicitly deal with milling tolerance. Their definition could, of course, be extended to include solids at various tolerance limits but such an extension would drastically increase the size of the problem (i.e. the number of Boolean operations) to be solved. In contrast, the definition presented in this dissertation deals with direct comparison of the desired part with the action of the mill; this allows the straightforward incorporation of a tolerance band. The two definitions theoretically agree in classifying the part as missed or gouged in the ideal case in which the tolerance band is set to zero. Of course, this comparison is possible only when the parts to be verified are defined as complete solids, which is not a requirement of the technique presented here.

N/C geometric verification severity problem

A useful extension to N/C geometric verification involves consideration of the degree or magnitude of the deviations (misses and/or gouges) of the tool path from the desired part model. It involves first defining a "range of interest" which bounds the maximum miss or gouge which is to be portrayed distinctly from less severe ones; that is, more severe misses or gouges will be "lumped" as severe misses or gouges and further degree of severity information will be lost. Using the same terminology as the above formulation, the N/C geometric verification severity problem is defined as follows. Let,

R_{int} denote a range of interest which designates the gouge,

$L_{out} = R_{int} + T_{out}$, be the cut value limit for the maximum distinguishable miss, and,

$L_{in} = R_{int} + T_{in}$, be the cut value limit for the maximum distinguishable gouge.

Define a <u>degree of cut</u> mapping,

$D : P \to \{ -1, G(p_i), 0, M(p_i), 1 \}$, such that,

$$D(p_i) = \begin{cases} -1 & \text{if } C(p_i) \leq -L_{in} \\ G(c_i) & \text{if } -L_{in} < C(p_i) < -T_{in} \\ 0 & \text{if } -T_{in} \leq C(p_i) \leq T_{out} \\ M(c_i) & \text{if } T_{out} < C(p_i) < L_{out} \\ 1 & \text{if } L_{out} \leq C(p_i) \end{cases}$$

where, $G(p_i)$ and $M(p_i)$ are functions of the cut value, for example,

$G(p_i) = (C(p_i) + T_{in}) / R_{int}$, and
$M(p_i) = (C(p_i) - T_{out}) / R_{int}$.

The program N is <u>geometrically verified</u> iff for all $p_i \in P$, $D(p_i) = 0$, and for all $p_i \in H$, $E(p_i) = 0$. P is <u>gouged</u> if there exists p_i such that $D(p_i) = G(p_i)$ and <u>gouged severely</u> if there exists p_i such that $D(p_i) = -1$. Similarly, P is <u>missed</u> if there exists p_i such that $D(p_i) = M(p_i)$ and <u>missed severely</u> if there exists p_i such that $D(p_i) = 1$.

The definition of holding fixture interference is the same as the one described above for direct N/C geometric verification. Also, the regions of miss and gouge are constructed in an analogous manner except that cut values are interpolated according to an interpolation rule. This rule incorporates information concerning the magnitude of the deviations between the as-milled part and the desired part. The linear interpolation rule presented above is only an example, it could be replaced with other functions, based perhaps on tool geometry or a logarithmic function of the cut value.

Note that this definition of the "severity" problem reduces to the more simple N/C geometric verification problem if the range of interest, R_{int}, is set equal to zero. Another interesting case occurs if the tolerance band is set to zero ($T_{in} = T_{out} = 0$) and the R_{int} is set equal to the tool radius. Under these conditions, very detailed features of the surface (as affected by the mill) would be highlighted, e.g. cutter cusps.

This extended definition of the N/C geometric verification problem is even more distinct from the direct solid modeling problem definition. The solid modeling approach results in mathematical models resulting from discrepancies between the tool path and the desired part. However, the solid modeling problem definition does not explicitly address the quantification of the degree of discrepancy which is represented by these solids. In contrast, incorporation of the "degree of" miss and gouge information into the problem definition described in this work is a simple and natural extension, and can be used to illustrate graphically the severity of milling problems.

ALGORITHMIC IMPLEMENTATION

The N/C geometric verification problems described above are easily discretized to allow an algorithmic implementation for their approximate solution. Such an algorithm has been developed and tested with excellent results (Oliver[7]). However, a detailed description of the algorithm is beyond the scope of this paper. The following discussion serves as a general overview.

The algorithm for direct N/C geometric verification developed by Oliver[7] is a new application of the technology underlying surface and solid modeling; it combines techniques originally developed for accurate non-polygonal surface shading and elements of B-rep solid modeling. The algorithm is designed with the goal of minimizing unnecessary computations. It involves checking tool paths against the desired part geometry (and fixturing, if necessary) to produce graphical output depicting the desired part as shaded surfaces with out-of-tolerance areas highlighted. The current implementation is applicable to three-axis milling operations.

The algorithm is most easily described in three phases: a preprocessing phase, in which the surfaces of the workpiece model are discretized into points and normal vectors, a tool path processing phase, in which cut values are calculated for each point, and a postprocessing phase, in which cut value and normal vector are used to assign a hue and intensity to each point.

Preprocessing
Rational B-spline surfaces (uniform or nonuniform) are used to model the workpiece, which may include holding fixtures. An efficient scanline sculptured surface shading algorithm (Pickelmann[8] and Vanderploeg[9]) is applied in the preprocessing phase to yield an accurate surface coordinate and normal vector at each pixel (picture element) of a raster-type display device which maps onto the workpiece model.

The second step in preprocessing involves transformation of the pixel coordinate data into an orientation which facilitates efficient correlation of a mill motion with a specific area of the workpiece model. A linked list data structure is developed to allow processing in "mill axis space" the coordinate data generated in "view space".

Tool path processing
Tool path processing is the heart of the N/C geometric verification algorithm and it is fundamentally independent of the method used for part model discretization. The mill path is modeled as a series of implicit B-rep solids which represent sequential volumes swept out by the mill. The tool path is processed in the two-dimensional mill axis space to determine a fairly tight bound on the subset of pixels which could possibly have been affected by a given motion of the mill. Then the linked list is used to access the three-dimesional view space coordinate and normal data to perform the actual intersection calculation. When all pixels in the subset have been considered, the swept volume is discarded and the next tool motion is considered.

The intersection calculation for any given normal vector proceeds in hierarchical steps only as far as needed through a series of progressively more exact definitions of the shape of the tool swept volume. The results of intermediate calculations are used to determine if further, more sophisticated swept volume intersection calculations are required. This structure insures that redundant or superfluous calculation of vector/solid intersection is minimized.

If, during processing, a normal intersection is calculated yielding a cut value which is less than the one previously stored for that pixel, the new value is saved, overwriting the previous value. Thus on completion of processing, only the deepest excursion of the mill toward or into the surface is saved for each pixel.

Postprocessing
The output image of the sculptured surface part, as milled by the tool path, is generated using the cut value and normal vector saved at each pixel. The color of each pixel is made up of a hue and an intensity. Pixel hue is determined by the cut value. When the range of interest is set to zero, green represents a cut within tolerance; red, a gouge; and blue, a miss. For a nonzero range of interest (i.e., addressing the severity problem), hue interpolation is applied to convey the "degree of" miss and gouge information. In this case a gouge varies from red for a slight gouge through yellow for the maximum. Similarly, misses range from blue for the slight miss case through magenta for the maximum. Independent of hue, pixel intensity is based on the angle that the normal vector makes with a user-selectable light source, thus enhancing surface feature recognition (Lambertian shading).

Cut value interpolation The algorithm allows the user to trade computational speed for accuracy (resolution) of verification through a parameter called the discretization element size. This parameter is applied in the preprocessing phase to reduce the number of normal vectors which participate in the precise intersection calculation. Let the discretization element size be denoted by M; it is applied by transforming only every Mth pixel on every Mth scanline into mill axis space. Cut values for pixels which are not transformed (i.e., their associated normals do not participate in swept volume intersection) are interpolated via a linear, nearest neighbor scheme, based on the distance from the pixel in question to the (up to) four closest pixels which were calculated exactly. Interpolation is only necessary, of course, when the discretization element size is set equal to a value greater than one. This has an effect similar to "zooming out" for the verification calculation without affecting the size of the output display.

RESULTS

Two application examples are now presented which demonstrate the capabilities of the direct N/C geometric verification algorithm described above. First, a portion of a simple manufactured part is considered, to illustrate the interpretation of the graphical output. This example also demonstrates verification of contour milling operations and the incorporation of holding fixtures as part of the workpiece model. In the second example, a part from the aerospace industry is considered. Several test runs are made illustrating the effects of the user-selectable parameters (i.e., range of interest and discretization element size) on both resolution and overall computation time. Tradeoffs between the conflicting goals of rapid computation time and high resolution verification are discussed, as are the limitations of the user-selectable parameters.

Both application examples were run on a Prime 750 superminicomputer with 6 megabytes of physical memory; results were displayed on a Tektronix 4129 color terminal. In all of the application examples the computation time for the preprocessing phase of the algorithm, i.e., calculation of surface point coordinates and normals, is reported separately from the computation time required for N/C geometric verification. This is done because preprocessing time is a fixed cost (independent of the size of the CL-data file) and the verification procedure is essentially independent of the method used for surface discretization and generation of normals. For example, a more traditional polygon tiling of the surface could be used, resulting in faster computation time due to the simpler calculation of (generally fewer) normal vectors.

Contouring operation
Although it could represent an actual manufactured part, the part model (and associated N/C program) considered in the first application example was created by the author for demonstration purposes. This example features N/C geometric verification of a contouring operation on a workpiece model that includes two holding fixtures. The part is essentially a three inch thick slab of material with one corner rounded into a three inch radius cylindrical surface. The part model (including holding fixtures) consists of 14 rational B-spline surfaces; the associated CL-data file contains only 19 points, in order to illustrate milling errors. A spherical end tool of radius 0.5 inch and height 3.0 inches was used to mill the contour. The user selectable parameters were set as follows: discretization element size equal to one, milling tolerances equal to 0.01 inch (inside and outside), and range of interest equal to 0.49 inch.

Results of the N/C geometric verification algorithm as applied to this model are shown in Figure 1. Since the N/C program only mills (contours) the sides of the part, the entire top of the slab appears as missed by at least the maximum amount, 0.5 inch. The walls or sides of the part appear to be

114 NUMERICAL CONTROL PROGRAMS

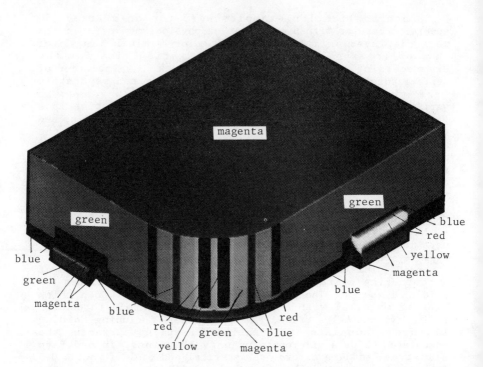

Figure 1. N/C geometric verification of a contouring operation

milled within tolerance except in the area of the cylindrical surface where both slightly missed and gouged regions are apparent. In this example, the circular arc tool path necessary to mill this area precisely was approximated by eight linear segments. The errors were intentionally introduced to show the verification capability. The band, ranging from blue to magenta, around the bottom of the part shows the effect of milling a corner with a ball end tool.

Figure 1 also shows the effect of the mill on holding fixtures. Both fixtures extend 0.5 inch out from the sides of the part. The fixture on the lefthand side of the part measures 0.5 inch in height, while the one on the right has a height of 1.0 inch. The mill successfully avoids the fixture on the lefthand side since it is shaded in blue and magenta with only a small green area that was affected within the tolerance limit. However, on the righthand side of the part, the tool interferes with the fixture. Note that the fixture is shaded as modeled (a planar box), but that its yellow hue indicates a severe gouge.

The computation time required for this application example was approximately 25 CPU minutes for the preprocessing phase (the calculations needed for ordinary shading of the image) and 19 CPU minutes for N/C geometric verification.

Turbine blade
The second N/C geometric verification example considers a faulty part program (generated by N/C software which was then under development) which was used to machine an actual prototype. The desired part in this example is modeled by a single rational B-spline surface patch representing the convex side of a turbine blade. The part is approximately 1.75 inches wide by 3.5 inches long. It is to be milled (on a three-axis milling machine) with a ball end tool of radius 0.25 inch and height 1.0 inch. The CL-data for this part consists of 3034 distinct positions of the mill tool. In each of the example runs, the milling tolerance was $T_{in} = T_{out} = 0.001$ inch.

Results of the N/C geometric verification of this N/C program are shown in Figures 2, 3 and 4; associated computation times are given in Table 1. Figure 2 shows the results of the most accurate, and hence, the most computationally intensive run. In this case, the discretization element size was set to 1 (i.e. every pixel was considered in the intersection calculation) and the range of interest was set to 0.01 inch. Figure 3 shows results with the same conditions as above, but with the range of interest set to zero. Finally the results of the same part in the same view, with discretization element size set to 4 and range of interest set to 0.01 inch, are shown in Figure 4.

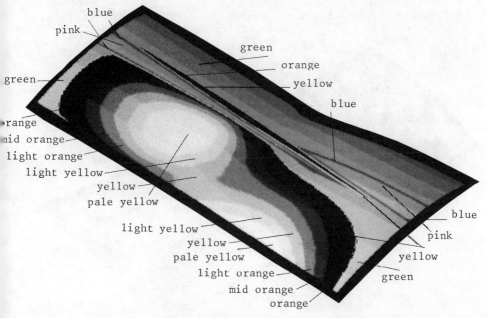

Figure 2. Turbine blade N/C geometric verification, range of interest = 0.01 in., discretization element size = 1

116 NUMERICAL CONTROL PROGRAMS

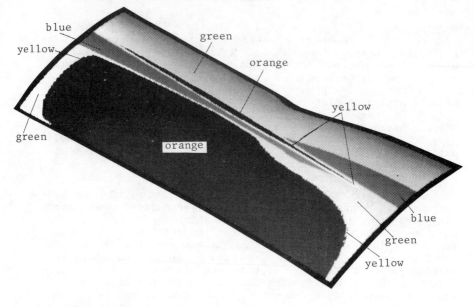

Figure 3. Turbine blade N/C geometric verification, range of interest = 0.0 in., discretization element size = 1

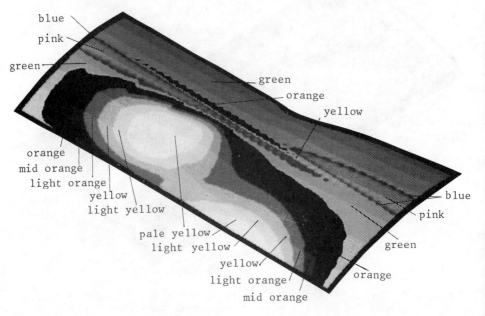

Figure 4. Turbine blade N/C geometric verification, range of interest = 0.01 in., discretization element size = 4

It should be noted that the "color banding" apparent in Figures 2 and 4 is a result of the limitations of the display terminal used for this work. The terminal is limited to display at most 256 colors at a time. When the user selects a nonzero range of interest, 15 hues are used to depict depth of cut (7 each for miss and gouge and one for "in tolerance"). The color limitation results in only 16 intensity levels being available for each hue. The result shown in Figure 3 appears smoother because only three hues are used if the user selects a range of interest equal to zero. For this case, there are 80 intensity levels available for each hue.

Table 1 summarizes the results of these three cases and breaks the total computation time down by function. In each case the total CPU time spent to calculate the surface normal vectors was about 21 minutes.

These results show the dramatic effect that the discretization element size has on the performance of the algorithm. Computation time was reduced by a factor of approximately 8 when the discretization element size was changed from 1 to 4. Of course, this change has an adverse effect on the resolution and accuracy of the resulting output image, but it is still quite apparent that the improper milling of the part would be discovered in this case. An even greater

Table 1

Computation Time for N/C Geometric Verification
Turbine Blade Example

	Computation time CPU minutes
Figure 2. Range of interest = 0.01 in. Discretization element size = 1	404.0
Figure 3. Range of interest = 0.0 in. Discretization element size = 1	395.7
Figure 4. Range of interest = 0.01 in. Discretization element size = 4	51.2

reduction in computing time can be obtained simply by "zoomimg out" on the part, at the cost of a smaller resulting image and coarser resolution.

Reducing the range of interest also cuts computation time although not nearly as dramatically as the discretization element size change. This example shows that the computational price for the "degree of" miss or gouge information is really quite small. The effect of the discretization element size is much more profound since it actually reduces the size of the data set (pixels and normals) with which the tool path must be compared.

These example applications show how the user may apply variations in the range of interest and discretization element size, along with prudent view selection, to increase computational efficiency. Alternatively, if he is prepared to pay the computational price, the user may opt for precision limited only by the resolution of the display device.

DISCUSSION

An order of complexity analysis presented by Oliver[7], shows that the N/C geometric verification algorithm described here behaves with a time complexity of $O(N)$, where N is the cardinality of the CL-data file. This compares with results of a similar analysis by Voelcker and Hunt[3] which show that a direct solid modeling approach to N/C geometric verification has time complexity in the general case of $O(N^4)$ and $O(N^3 \log(N))$ under certain special circumstances. This represents a dramatic difference in computational effort.

More recent research efforts have dealt with improving the computational efficiency of the solid modeling approach. In particular, Wang[10,11] combines a scanline algorithm with image space Boolean operations and a mathematical basis for full five-axis tool swept volumes (which includes the rotational speed of the mill) to perform N/C geometric verification with impressive computational speed. Van Hook[12] presents a method which applies image space Boolean operations on volume (depth) elements, rather than sightlines, in a Z-buffer algorithm which provides real-time N/C simulation. He also suggests extensions to the technique necessary for N/C geometric verification.

A surface-based approach, similar in some respects to the technique presented here, was recently proposed by Jerard[13]. Jerard applies a first order curvature approximation to evaluate a grid of points on the surface which is dense enough to account for surface irregularities relative to the size of the mill tool. He does not evaluate surface normals but instead applies a Z-buffer algorithm on the grid itself as he evaluates the tool paths. Jerard's technique is limited to three-axis milling; the surface "cut values" are measured relative to the mill axis.

The diversity of techniques presented in this paper and the efforts mentioned above suggest that the general problem of automated N/C verification is far from solved. Issues which require further research include full five-axis and sculptured surface capabilities, combined simulation and geometric verification, improved computational efficiency and accuracy, and simple user interfaces. These authors are also investigating the synthesis of tool path generation and verification to improve part programming productivity.

REFERENCES

1. Smith D.A. and Evans E. (1977). Management Standards for Computer and Numerical Control, Institute of Science and Technology, Industrial Development Division, Ann Arbor, MI.

2. Voelcker H.B. and Hunt W.A. (1981) The Role of Solid Modelling in Machine-Process Modelling and NC Verification, Proceedings of SAE 1981 International Congress and Exposition, Detroit, MI.

3. Hunt W.A. and Voelcker H.B. (1982) An Exploratory Study of Automatic Verification of Programs for Numerically Controlled Machine Tools, Production Automation Project, Technical Memo No. 34, University of Rochester.

4. Ruberl S.T. (1981) Verification of NC Part Programs with Interactive Computer Graphics and Solid Geometric Modeling, Proceedings of CAM-I 10th Annual Meeting and CAD/CAM Graphics User's Exposition, Fort Worth, TX.

5. Fridshal R., Cheng K.P., et al (1982) Numerical Control Part Program Verification System, Proceedings of the Conference on CAD/CAM Technology in Mechanical Engineering, Massachusetts Institute of Technology.

6. ANSI (1982). Dimensioning and Tolerancing, ANSI Y14.5M-1982, ASME, New York, NY.

7. Oliver J.H. (1986). Graphical Verification of Numerically Controlled Milling Programs for Sculptured Surface Parts, Ph.D. Dissertation, Michigan State University.

8. Pickelmann M.N. (1985). The Design of Rational B-Spline Algorithms for Interactive Color Shading of Surfaces, Ph.D. Dissertation, Michigan State University.

9. Vanderploeg M.J. (1982). Surface Assessment Using Color Graphics, Ph.D. Dissertation, Michigan State University.

10. Wang W.P. (1984). Solid Geometric Modeling for Mold Design and Manufacture, Ph.D. Dissertation, Cornell University.

11. Wang W.P. (1985) Integration of Solid Geometric Modeling for Computerized Process Planning, Computer-Aided / Intelligent Planning - PED, Vol. 19, Book No. G00334, ASME, New York, NY.

12. Van Hook T. (1986) Real-Time Shaded NC Milling Display, Computer Graphics, Proceedings of SIGGRAPH, Vol. 20.

13. Jerard R.B. Hauck K. and Drysdale J.K. (1986) Simulation of Numerical Control Machining of Sculptured Surfaces, Proceedings of the International Symposium on Automotive Technology and Automation, Flims, Switzerland.

A Vibratory Device for Locating Objects: Theory and Experimental Results

D.T. Pham and J. Menéndez
Department of Mechanical Engineering, The University of Birmingham, Birmingham B15 2TT, United Kingdom

ABSTRACT

This paper describes a novel sensor to determine the location of rigid objects. The device can be mounted on the end of a robot arm to enable it to sense the position and orientation of a part it has picked up from a semi-ordered stack. It can also be used in a stand-alone mode to measure the co-ordinates of components delivered to it by a conveyor or a chute.

The device consists of a platform-motor assembly able to oscillate about an axis normal to that of the motor shaft. The object to be located rests on the platform and is rotated to two positions by the motor. At each of these, the static displacement of the platform and the frequency of vibration about the axis of oscillation are measured.

The mathematical procedure for computing the co-ordinates of the object relative to the platform from these measurements is described. Experimental results obtained with a test rig which simulates the device are presented.

INTRODUCTION

In the automotive industry, the need often arises to handle reasonably heavy parts, that is, those weighing between 200 g. and 2000 g. A problem with robotising the handling of such parts is that vibratory equipment, conventionally used to supply small items in an orderly fashion to robots, is not suitable.

This paper describes a sensing device which a robot could use to feed itself with the type of parts mentioned above. The device belongs to a family of vibratory sensors investigated by the authors's group (Pham and Dissanayake[1,2,3]) over the past few years. As has been the case with these sensors, it is assumed that parts are initially "semi-ordered", their position and orientation being known to ±25 mm and ±45° respectively.

After summarising the mechanical features of the proposed sensor, the paper will present its underlying theory and the results of experiments to demonstrate its feasibility. The paper concludes by outlining other modes of operation of the sensor in which the requirement for semi-ordering parts is removed by giving the robot additional sensing capability.

DESCRIPTION OF SENSOR

Figure 1 is the schematic diagram of a rig simulating the proposed sensor. The platform (P) can be rotated about the axis $\underline{Z}$ and positioned at any required angle using the stepping motor (S). The torsional spring (T) with stiffness K restrains the vibration of the platform-motor assembly about the axis AA. Static deflections are measured using a Linear Variable Differential Transformer (LVDT). An initial impulse to set the system into free vibration is provided using the solenoid (So). The resulting motion can then be picked up using the same LVDT. This signal can be processed to obtain the period of vibration. Note that a magnetic sheet (MS) is on top of P in order to hold the object firmly with respect to P. This limits the use of the sensor to ferrous objects but by resorting to other holding means the range of objects can be widened.

THEORY

It is assumed that the object lies flat on the platform and can be completely located if one knows the coordinates (X_G, Y_G) of its centre of mass G and the angle γ between one of its reference axes $(\underline{X}')$ which is parallel to P and a coordinate axis $(\underline{X})$ in the plane of P. By measuring static deflections at two different positions of the platform (X_G, Y_G) can be found. γ can be calculated from measurements of the period of vibration at those positions. Figure 2 summarises the steps in the location of an object using the proposed device.

<u>Determination of (X_G, Y_G)</u>
If the static deflection of P about AA is small and equal to θ_{s1}, the mass of O is m_0, and the stiffness of T with respect to AA is K and supposing that P is horizontal when there is nothing on it, X_G can be obtained from the following static equilibrium equation

$$K\theta_{s1} = m_0 g (X_G - E) \qquad (1)$$

where E is the distance between AA and the axis of S. Y_G is computed from

$$K\theta_{s2} = m_0 g ((X_G - Y_G) \cos 45° - E) \qquad (2)$$

where θ_{s2} is the static deflection about AA of P after S has rotated it through 45°.

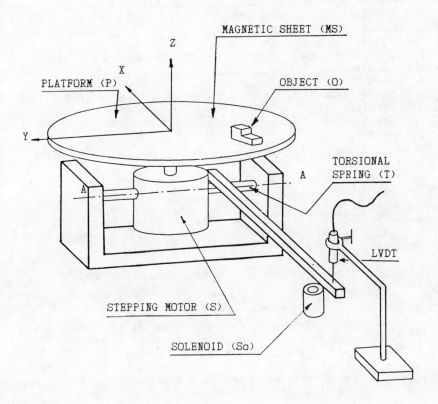

Figure 1. Proposed sensor

Determination of orientation γ

The system has one degree of freedom and its natural frequency (ignoring damping) is given by

$$\omega^2 = (K - M_p g Z_p - m_o g Z)/I_{AA} \tag{3}$$

where
ω = Natural frequency of the platform-object system.
K = Stiffness of the torsional spring.
M_p = Mass of the platform.
g = Gravitational constant.
Z_p = Height from the centre of mass of the platform to the axis of vibration.
m_o = Mass of the object.
Z = Height from the centre of mass of the object to the axis of vibration.
I_{AA} = Inertia of the platform-object system about the axis of vibration.

The inertia of the platform-object system can be written as

$$I_{AA} = I_p + I_o \tag{4}$$

where I_p and I_o are the inertias of the platform and object respectively. On the other hand I_o can be expressed as follows:

$$I_o = (I_{XX} + I_{YY})/2 + [(I_{XX} - I_{YY}) * \cos 2\gamma]/2 - I_{XY}\sin 2\gamma + m_o(Z^2 + D^2) \tag{5}$$

where
I_{XX}, I_{YY} = Moments of inertia of the object with respect to axes X' and Y' fixed to its centre of mass G. I_{XX} and I_{YY} are known for a given object.
I_{XY} = Product of inertia. I_{XY} is known for a given object.
γ = Angle between the axis X fixed on the platform and the axis X' fixed in the object. See Figure 3.
D = Distance along the X axis of the platform from the centre of mass of the object to the axis of vibration on the platform. D is known when X_G is determined.

The period of vibration is

$$T = 2\pi/\omega \tag{6}$$

Then from Equations 3, 4 and 6

$$I_o = \frac{1}{4\pi^2}(K - M_p g Z_p - m_o g Z)T^2 - I_p \tag{7}$$

The values of K, M_p, Z_p, I_p are constant and can be found

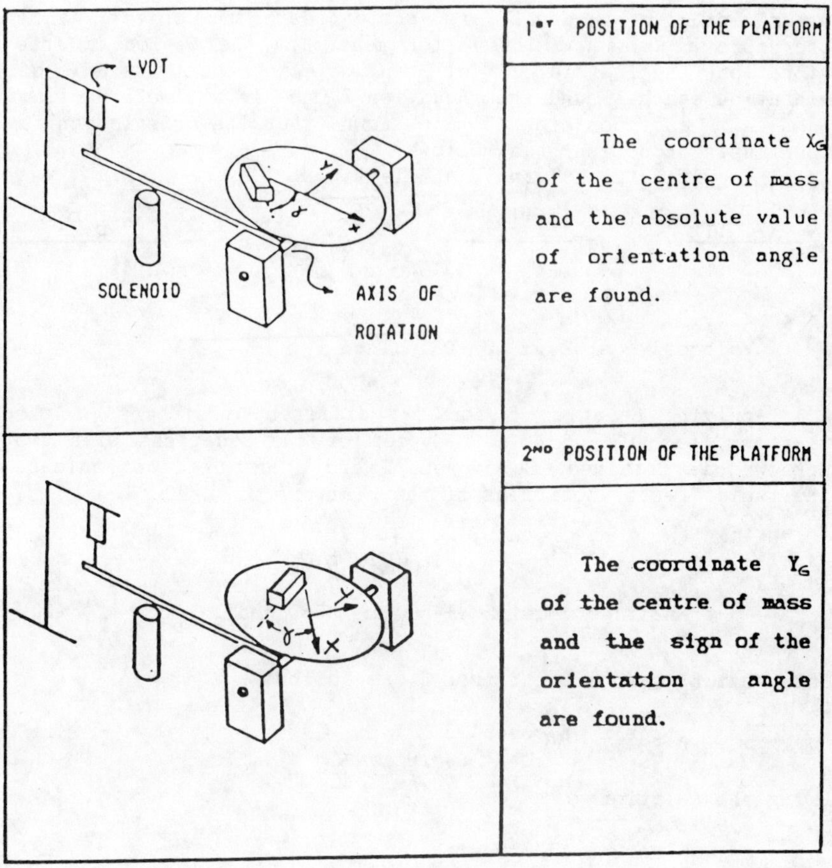

VIBRATORY DEVICE FOR LOCATING OBJECTS 125

	1ˢᵗ POSITION OF THE PLATFORM
(diagram: LVDT, SOLENOID, AXIS OF ROTATION)	The coordinate X_G of the centre of mass and the absolute value of orientation angle are found.
(diagram)	2ⁿᵈ POSITION OF THE PLATFORM
	The coordinate Y_G of the centre of mass and the sign of the orientation angle are found.

Figure 2. Steps in locating an object

126 VIBRATORY DEVICE FOR LOCATING OBJECTS

experimentally. m_o and Z are properties of the object and therefore also known. Thus after measuring the period of free vibration, the inertia I_o of an object about the axis of vibration can be found from Equation 7. Once I_o as well as I_{XX}, I_{YY}, I_{XY}, and position D are known then the orientation is found using Equation 5 as follows.
Combining Equations 5 and 7 and letting

$$a = \frac{I_{XX}+I_{YY}}{2} \qquad b = \frac{I_{XX}-I_{YY}}{2} \qquad c = I_{XY} \qquad d = \frac{K - M_p g Z_p - m_o g Z}{4\pi^2}$$

yield :-

$$a + b\cos 2\gamma - c\sin 2\gamma + m_o(Z^2 + D^2) = dT^2 - I_p \qquad (8)$$

Applying Equation 8 for two different positions of the platform, say 0° and 45°, a system of two equations with two unknowns is obtained. Subscripts 1 and 2 are used to indicate the two different positions of the platform.

$$a + b\cos 2\gamma - c\sin 2\gamma + m_o(Z^2 + D_1^2) = dT_1^2 - I_p \qquad (9)$$

$$a + b\cos 2(\gamma+45°) - c\sin 2(\gamma+45°) + m_o(Z^2 + D_2^2) = dT_2^2 - I_p \qquad (10)$$

Note that D_2 can be found from D_1 as follows

$$D_2 = (D_1 - E)\cos 45° + E \qquad (11)$$

Using the relations

$$\cos 2(\gamma+45°) = -\sin 2\gamma \qquad (12)$$
$$\sin 2(\gamma+45°) = \cos 2\gamma \qquad (13)$$

and rearranging Equations 9 and 12 produce :-

$$b\cos 2\gamma - c\sin 2\gamma = dT_1^2 - I_p - m_o(Z^2 + D_1^2) - a \qquad (14)$$

$$-b\sin 2\gamma - c\cos 2\gamma = dT_2^2 - I_p - m_o(Z^2 + D_2^2) - a \qquad (15)$$

Equations 14 and 15 can be solved to give :-

$$\sin 2\gamma = \frac{c(dT_1^2 - I_p - m_o(Z^2 + D_1^2) - a) + b(dT_2^2 - I_p - m_o(Z^2 + D_2^2) - a)}{-b^2 - c^2} \qquad (16)$$

$$\cos 2\gamma = \frac{-b(dT_1^2 - I_p - m_o(Z^2 + D_1^2) - a) + c(dT_2^2 - I_p - m_o(Z^2 + D_2^2) - a)}{-b^2 - c^2} \qquad (17)$$

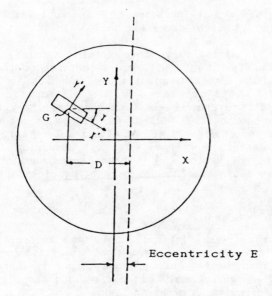

PLAN

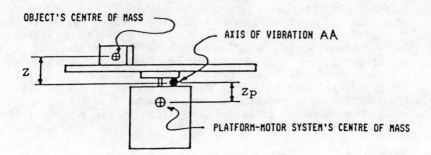

ELEVATION

Figure 3. Coordinate systems and position of the centre of mass

128   VIBRATORY DEVICE FOR LOCATING OBJECTS

Clearly, there are two angles between 0° and 360° which satisfy Equations 16 and 17 which are γ and γ+180°.

It is easy qualitatively to explain the above results. An object as in Figure 4, with known position of the centre of mass, has four possible orientations in which the inertia around an axis AA is the same. If the platform is rotated through, say 45°, to another position then the inertia of the object in this new position will reveal if the object was like in Figures 4a, 4b or Figures 4c, 4d. The indeterminacy between γ and γ+180° cannot be resolved by just measuring inertias. However this is not serious as it has been assumed that γ is known to within ±45° prior to the part being picked up by the robot.

CONTROL OF SENSOR

A control programme was written in Basic on a Motorola Exorset microcomputer to move the platform to specified positions, excite it and measure the period of vibration using three M6809 assembly language subprogrammes which interface the computer to the electronic circuits as shown in Figure 5.

EXPERIMENTAL RESULTS

A set of parallelepipedic objects was used in order to find how different ratios of width/length would affect the accuracy in orientation γ. Also objects of the same width/length ratio but different masses were studied so that conclusions about accuracies in positioning G and finding γ could be drawn. The dimensions and masses of the objects are given in Table 1. Table 2 shows where the objects were actually placed. All the measurements were done after the temperature of the mechanical hardware was stable. Usually, this required waiting for half-an-hour after first switching the apparatus on. The accuracy in the location as a function of the distance from the centre of mass of objects of different dimensions to the centre of the platform is shown in Figures 6 and 7.

DISCUSSION OF RESULTS

The accuracy of the position of the centre of mass is better for the heavier objects in the samples. Also it is improved for greater distances away from the centre of the platform. These two factors can be summarised as: the greater the torque applied to the torsion bar, the better the accuracy of the position of the object's centre of mass. This was expected because a difference of 1 mm in the position when the torque is high produces a larger difference in platform deflection than if the torque is low. This larger difference in deflection is more easily detected by the LVDT.

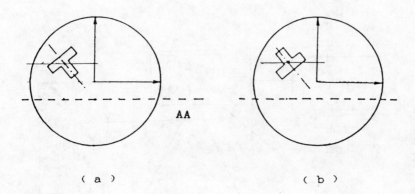
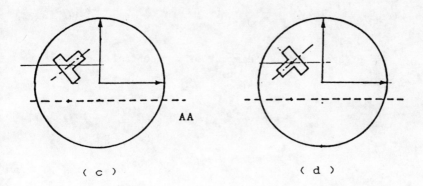

Figure 4. Indeterminacy in the orientation

130 VIBRATORY DEVICE FOR LOCATING OBJECTS

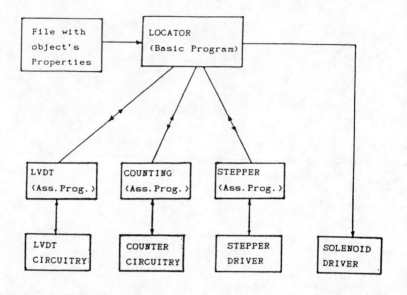

Figure 5. Block diagram of the control system

OBJECT	LENGTH [mm]	WIDTH [mm]	HEIGHT [mm]	WIDTH/LENGTH	MASS [gr]
A1	100.0	30.0	5.0	0.3	117.0
A2	100.0	30.0	10.0	0.3	235.0
A3	100.0	30.0	15.0	0.3	352.0
B1	65.6	45.9	5.0	0.7	117.0
B2	65.6	45.9	10.0	0.7	235.0
B3	65.6	45.9	15.0	0.7	352.0

Table 1. Sample data.

COORDINATES OF THE CENTRE OF MASS		ORIENTATION
X (mm)	Y (mm)	(deg)
0	0	30
7	7	30
14	14	30
21	21	30

Table 2. Actual location of the samples.

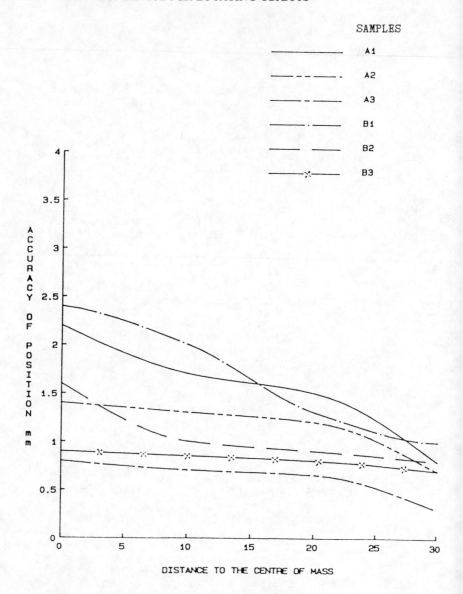

Figure 6. Accuracies in the location of centre of mass

VIBRATORY DEVICE FOR LOCATING OBJECTS 133

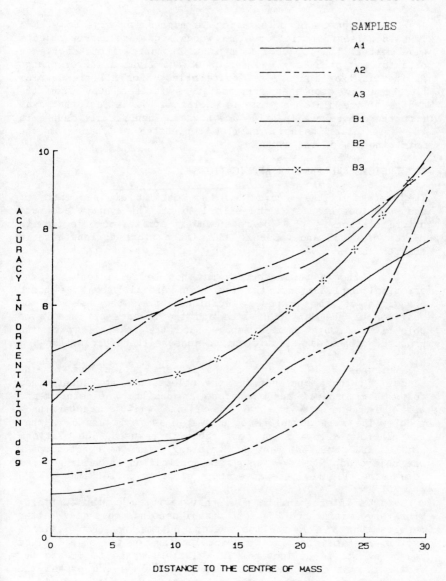

Figure 7. Accuracies in the orientation γ

The accuracy of orientation is higher for objects with a low width/length ratio. In other words, long and narrow objects give better results. This is because a change of one degree in orientation for a long and narrow object gives a larger change in inertia than for a more square object. For all the samples the accuracy decreases for positions further away from the centre of the platform. This is explained by the fact that away from the centre the inertias are higher. Then a small change in the orientation means a smaller percentage of change in the total inertia of the system.

CONCLUSIONS AND FURTHER APPLICATIONS

The average of the accuracy of the position of the centre of mass considering all samples is 1.25 mm. The average accuracy of orientation is 2.0 degrees and it applies to objects with masses of 0.235 and 0.352 kg and with a width to length ratio smaller than 0.5 .

Even though this accuracy may not be sufficient in some practical applications, the tests in general prove the basic principle of the proposed technique, i.e. the use of the position of the center of mass and the moments of inertia of objects to locate them in space, and the feasibility of the proposed mechanical arrangement to measure these parameters.

Figures 8, 9 and 10 show further applications of the sensor. In these applications the objects are not constrained to be semi-ordered and therefore an auxiliary sensing means will be needed to resolve the γ and $\gamma+180°$ indeterminacy problem. The main advantage of a system as in Figure 8 is that the platform allows a low cost robot to be used. The platform rotates and gives the best possible angle for the robot to pick the object up. Figure 9 shows an application in feeding parts to a lathe. Figure 10 is an example of parts palletising.

It is interesting to note that having a system combining vibratory sensing and low cost visual sensing could give better results than any of the two systems on its own. For a vision system, most of the time is spent looking for the object. Thus measuring static deflections is a faster method to locate the centre of mass. On the other hand, with a vision system the above mentioned indeterminacy problem can be easily handled.

ACKNOWLEDGEMENTS

The authors wish to record their gratitude to their late Head of Department, Professor S.A. Tobias, for supporting this work. They would also like to thank Mr. P.L.M. Hurd of Birmingham University and Dr. R. Cox of the Research Corporation for organising the commercial exploitation of the sensor and the Committee of Vice-Chancellors and Principals for giving an ORS award to J. Menéndez.

REFERENCES

1. Pham D.T. and Dissanayake M.W.M.G. (1985) Feasibility Study of a Vibratory Sensor for Locating 3-D Objects. Proceedings of the 25 th Int. Machine Tool Design and Research Conf., Birmingham, United Kingdom, April 1985.

2. Pham D.T. and Dissanayake M.W.M.G. (1985) A Three Degree-of-Freedom Inertial Sensor for Locating Parts. Proceedings of the 15 th Int. Symp. on Ind. Robots, Tokyo, Japan, Sept. 1985.

3. Pham D.T. and Dissanayake M.W.M.G. (1985) Inertia-Based methods for Locating 3-D Objects. Proceedings of the 5 th Int. Conf. on Robot Vision and Sensory Control, Amsterdam, The Netherlands, Oct. 1985.

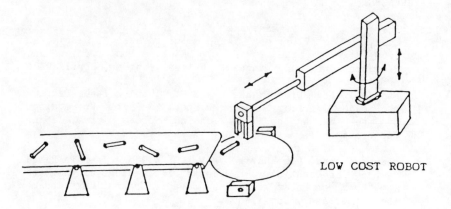

Figure 8. General application

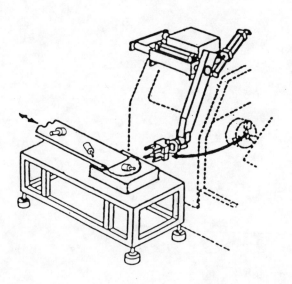

Figure 9. Lathe feeding application

VIBRATORY DEVICE FOR LOCATING OBJECTS 137

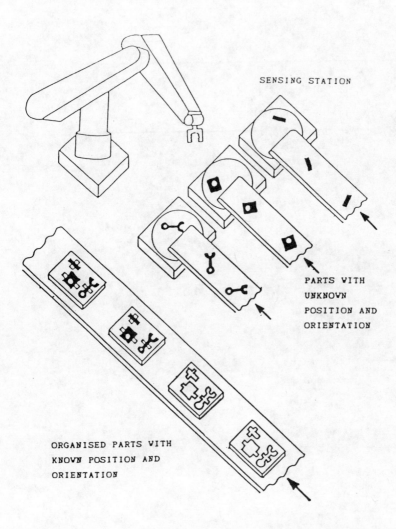

Figure 10. Palletising parts ready for assembly

SECTION 4 PERFORMANCE

Programmable Instrumentation Recorder and Support Computer for Racing Car Performance Testing

M.F. Bessant

Microsystems Design Group, Cranfield Impact Centre, Cranfield Institute of Technology, Cranfield, Bedford, England

ABSTRACT

This paper describes the design of an in-car recorder and track-side support computer system that is presently used by Formula One racing teams for chassis and engine performance testing. The latest C-MOS microcomputer devices were employed to produce a compact solid-state recorder that functions reliably, under severe environmental conditions, and can be programmed to accept signals from a wide range of instrumentation sensors.

Opto-isolation is included on the recorder's 8 digital and 3 pulse counting inputs, together with low-pass filtering for each of the 16 analogue channels. Although the recorder employs a 12-bit Analogue-to-Digital Converter, that offers a wide dynamic input range, data compression techniques ensure that only one (8-bit) byte of storage is required for each sample taken. Data storage is provided by 56,000 byte removable non-volatile RAM packs that can be rapidly exchanged during pit stops.

A low cost personal computer is used to support the recorder by interrogating the removable RAM pack and transferring the instrumentation data to disc. Graphics software was developed to enable data from individual channels to be rapidly searched and compared. Menu-driven software is also provided that enables an operator to select the rate at which each recorder input is sampled, together with the sensitivity and offset characteristics of individual analogue channels. The support computer is also used to program the recorders operating parameters into the RAM packs.

INTRODUCTION

Members of the C I C Microsystems Design Group have been collaborating with Formula One racing car teams for several years and have become familiar with their specialised performance testing procedures and instrumentation requirements. We soon found that one of the main operational constraints is associated with the expense of travelling to, and the hiring of, racing circuits. As the time spent at these test facilities is at a premium, instrumentation systems must be reliable, very easy to use and capable of rapidly displaying the effect of track-side engine and chassis modifications. In addition, all test data should be automatically labelled and stored in case it should be required for subsequent analysis.

When commissioned (by Formula One Race Car Engineering Ltd) to develop an instrumentation system, we decided that the programmable recorder, removable memory pack and integral support computer concept, outlined in Figure 1, would best meet the above requirements. Being inherently more reliable than traditional telemetry or tape based instrumentation systems and more flexible than a 'dumb' solid-state recorder.

The detailed design work was undertaken in the knowledge that the recorder/RAM pack combination would have to be mounted in an inhospitable area (e.g. in a radiator side-pod) due to the very limited space available in the car. Its electronics would thus be subjected to high vibration levels and wide temperature excursions. Being in close proximity to powerful ignition systems would also place it in a severe EMI environment, with fluid and dirt ingestion presenting an additional threat to reliabilty.

It was decided that the technology most suited to the above environment would be Complementary Metal Oxide on Silicon, by virtue of its wide operating temperature range and high level of noise immunity. In addition, CMOS has very low power dissipation which eliminates the need for ventilation and allows the electronics to be housed within a completely sealed (against EMI and fluid) enclosure. It is also used to fabricate LSI interface devices, the application of which has enabled us to minimise the number of component interconnections and produce a compact unit that is very tolerant of vibration and shock. The remainder of the paper will provide a detailed description of the instrumentation system that was used during the 1986 Formula One season.

PROGRAMMABLE INSTRUMENTATION RECORDER

The recorder, outlined in Figure 2, employs software programmable Versatile Interface Adapters to replace the many discrete logic devices (e.g. gates, monostables, latches, and counters) often associated with interfacing instrumentation signals to microcomputers.

The Rockwell 65C22 VIA (see Fig 3) device contains four flexible 8-bit counters, a serial -to- parallel/parallel -to- serial shift register and two latched parallel ports. Control of peripheral circuitry is handled primarily through the two 8-bit parallel ports. Each of these lines can be programmed as either an input or an output and each port has two additional I/O lines for handshaking. To facilitate control of this complex device, an interrupt register, an interrupt enable register and a pair of function control registers are also provided onboard.

Digital instrumentation sensors are obviously the type most readily acceptable to a VIA , requiring only the addition of opto-isolators for over voltage protection and noise immunity. However, even the 16 channel, 12-bit Analogue -to- Digital and pulse counting interface circuit (shown in Figure 4) uses very few components. Only the power rails and decoupling capacitors have been omitted from the diagram in the interests of clarity.

Potential users had indicated that the signals from analogue instrumentation sensors need not be measured to an accuracy of better than 0.4%, therefore each sample could be accommodated within a single byte of memory. In addition to offering the most efficient use of available storage, 8-bit measurements are also compatible with the support computers vertical resolution of 256 pixels. Unfortunately, a wide range of output voltages are produced by the different analogue sensors employed and an 8-bit ADC is inherently incapable of providing the necessary dynamic range. A 12-bit ADC was therefore used - offering a dynamic range of 74 dB - but for each input channel the Operating System consults a RAM based look-up table to find the relevant gain and offset settings (that had previously been programmed by the user) and strips off the unnecessary bits, until only a single byte of memory is needed for each sample taken. We therefore have a, low component count, software emulation of an 8-bit ADC preceded by programmable analogue gain and offset circuitry.

POWER SUPPLY

Instrumentation system wiring schemes have to be very carefully designed, if common-impedance path coupling and ground-loop induced noise problems are to be avoided. However, in motor racing the car and its instrumentation requirements are continually changing and the recording system must therefore be tolerant of sensor/battery wiring alterations. A high level of tolerance was achieved in the present design by employing digital opto-isolation techniques and an isolated DC-to-DC switchmode power supply.

The power supply draws approximately 250 mA from the cars battery and generates regulated system rails of +12, -12 and +5 volts. In addition, regulated +12 and -12 volt supplies are made available at each analogue connector for powering sensors. Separate current limiting (at 100mA) regulators for these outputs ensure that wiring and sensor faults are confined to individual channels.

REMOVABLE RAM PACK

Refering again to Figure 2, it can be seen that not only is the RAM removable but also the microproccesor and associated operating system EPROM. This may at first appear rather inefficient - even for the recorder's normal complement of two Packs - but this approach will simplify the introduction of larger storage capacity designs in the future. It also provides a convenient point in the system for inserting in-circuit emulators, during interface hardware and software development.

The present design of pack is based on a Rockwell 65C02 microprocessor that is capable of directly addressing a total of 64 Kbytes. The amount of RAM is restricted to 56K bytes in order to obtain an uncomplicated memory map, with the remaining 8K bytes being equally divided between I/O functions and the operating system firmware.

The RAM is implemented using seven pure C-MOS static memory devices (Toshiba 5564) that each draw approximately 45 mA during read/write operations and only 0.2 uA when in the standby mode. This semiconductor technology is thus ideal for non-volatile applications when supported by suitable retention circuitry.

MEMORY RETENTION CIRCUITRY

A rechargeable nickel-cadmium battery is included within each Pack to sustain the memory when it is not connected to the recorder or support computer. A fully charged battery is able to provide the few uA required by the C-MOS RAM for several months. However, the associated battery back-up circuitry has to be more complex than the simple diode networks often shown in I.C. and battery manufacturers literature, in order to perform the following two essential tasks.

The first task involves the memory devices. For C-MOS RAM to flawlessly retain data, in the absence of an adequate external supply, it must be placed in the standby mode and write-protected during power transients.

The second task involves control of the microprocessor. The processor is warned in advance of an impending power loss so that an orderly shutdown can be completed before the local reservoir capacitor voltage falls below tolerance. The processor is then forced into the reset state, which prevents any further bus activity that might disturb the dormant memory and corrupt data. When the power is restored the reset state is maintained until the processor clock has stabilized, the write-protect is removed and the memory has been switched back from the battery to the external supply.

OPERATING SYSTEM

The resident (in EPROM) Assembly Language Operating System software performs two main tasks, depending upon the state of a mechanical 'terminal present' link. If, at power-up, the support computer is physically connected to the recorders serial RS 232 port, the Operating System will enter the communications mode which supports real-time calibration and diagnostics tasks . Alternatively, if the link is absent, the operating system will run the data recording program which continually monitors the state of the cockpit mounted 'start recording' button until it is activated. The driver is presented with an LED display which is illuminated to indicate that the recorder is armed, flashed to confirm that data recording is in progress and is extinguished when the RAM pack is full.

SUPPORT COMPUTER

The support computer comprises of a single transportable unit (see Figure 6) that is capable of operating from a 12 volt battery. It employs an Acorn Master computer, with 128k RAM and dual 3.5 inch disk drives, together with a monochrome graphics display that has a resolution of 640 x 256. The unit also includes an interface that can transfer 56,000 bytes of data - from a RAM pack onto disk - within 40 seconds and can also be used to program the Pack with new operating parameters (eg. gain, offset and sampling rate for each channel) that the recorder must implement.

Although the main method of communicating with the Recorder is via the RAM packs , the serial interface was included so that real-time calibration and diagnostics could be performed over a detachable cable. This link can also be used to transfer data to disk, if the recorder's position on the car makes RAM pack removal difficult, but a complete transfer can take several minutes using this method.

The support computer -to- RAM pack interface (shown in Fig 5) provides another example of the level of circuit economy that can be achieved by using a programmable LSI device. The VIA based circuit is identical to that used by the Recorder for implementing the serial RS232 communication port and 8-bit parallel port.

In addition, both the VIAs are addressed at the same location in memory and can thus make use of the same resident communications software. When plugged into the support computer interface (and having first verified that the 'terminal link' is apparently present) a RAM pack will come under the control of the serial link, which will then initiate data transfer via the parallel port.

SUPPORT SOFTWARE

We have developed a menu-driven software package, for the Support Computer, that enables a user to select the Recorders operating parameters and also present the raw test data in graphical form. Two modes of screen presentation are available.

Using the dynamic mode it is possible to rapidly scroll through a single channel of data and identify particular events, such as brake pedal application. By selecting the static mode, data from any number of channels can be plotted and compared at the same point in the record. A hard copy of the resultant screen graphics may be requested from an Epson RX-80 dot-matrix printer. The examples of test results shown in Figure 7 were produced using this facility. Suspension travel is represented by channels 0, 1, 2 and 3, with brake pedal status shown on channel 9.

DATA ANALYSIS

The simplest method of transporting data is via the 3.5 inch disks but if the results of a track test are required urgently, at some remote development facility, data can be transmitted via the telecommunications modem provided. However, to date it has not been necessary to present the raw data to a more powerful computer as all analysis was performed using the resident BBC BASIC interpreter. This language has proved very popular in UK scientific and educational establishments, during the past few years, by virtue of its relatively fast structured implementation and convenient graphics handling qualities. It is therefore familiar to many racing car development personnel who have been able to create very specialised applications software.

CONCLUSIONS

Two systems, of the type described, have seen extensive service during the 1986 Formula One Season and proved to be very reliable. Operators found them to be powerful development tools, offering both flexibility and ease of use.

ACKNOWLEDGEMENTS

The author is grateful for the conceptual contributions made by Ross Brown of FORCE Ltd. and Martin Walters of Cosworth Engineering Ltd. I would also like to thank my colleague Jonathon Bloomer who was responsible for the Operating and Support systems software development.

PROGRAMMABLE INSTRUMENTATION RECORDER 149

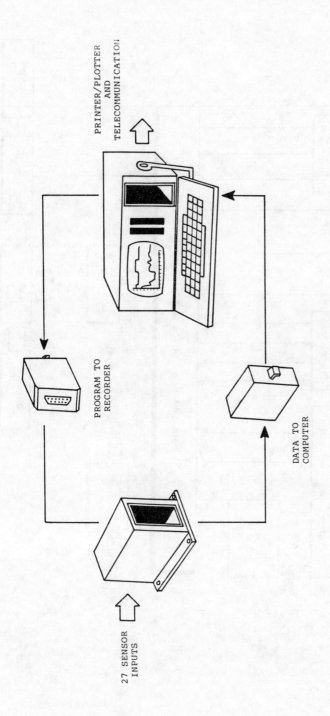

FIGURE 1. REMOVABLE RAM PACK CONCEPT

150 PROGRAMMABLE INSTRUMENTATION RECORDER

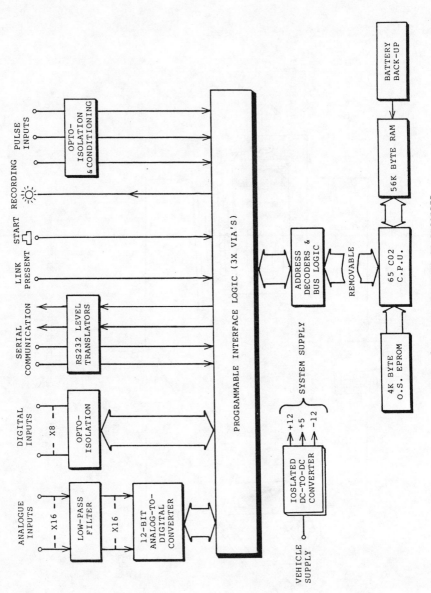

FIGURE 2. MAIN ELEMENTS OF RECORDER

PROGRAMMABLE INSTRUMENTATION RECORDER 151

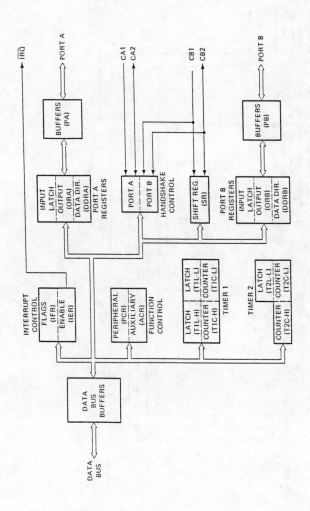

FIGURE 3. BLOCK DIAGRAM OF THE 65C22 VERSATILE INTERFACE ADAPTOR

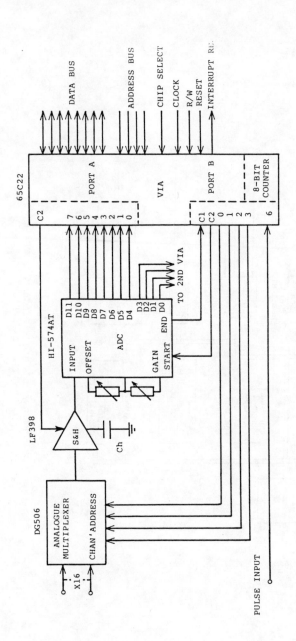

FIGURE 4. ANALOGUE AND PULSE COUNTER INTERFACE

PROGRAMMABLE INSTRUMENTATION RECORDER 153

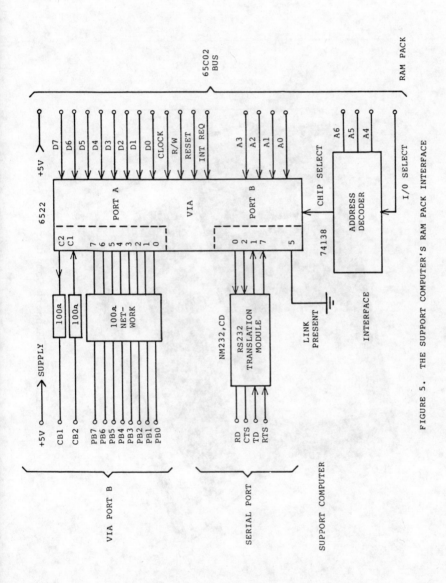

FIGURE 5. THE SUPPORT COMPUTER'S RAM PACK INTERFACE

Figure 6   Support Computer

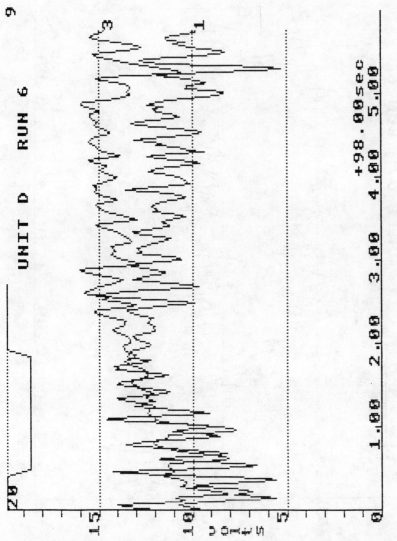

Figure 7a

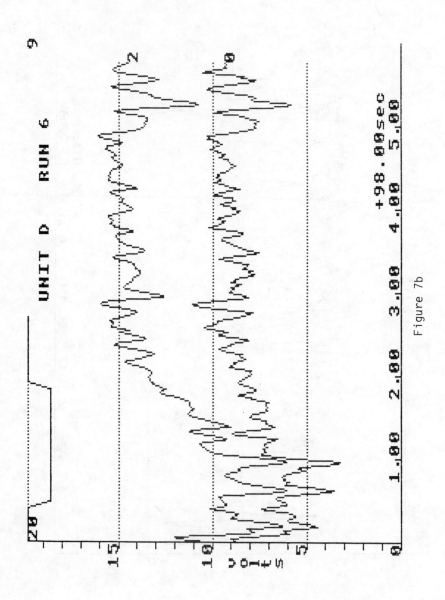

Figure 7b

PROGRAMMABLE INSTRUMENTATION RECORDER 157

Figure 8a   Ram Pack Electronics

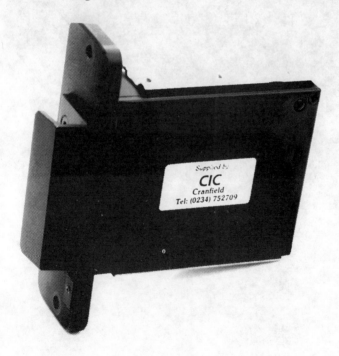

Figure 8b   Ram Pack

158 PROGRAMMABLE INSTRUMENTATION RECORDER

Figure 9   Programmable Recorder Unit

Route Simulation as a Product Development Tool
R.A. Buré
Manager Transport Consultancy Dept., Van Doorne's Bedrijfswagen Fabriek DAF B.V., Holland

INTRODUCTION

When developing commercial vehicles, it is of vital importance to have detailed knowledge of the actual operating conditions and subsequently to be able to test the function and lifespan of the vehicle design in these conditions.
Furthermore, constant endeavours are made to achieve further improvements in both the quality and efficiency of the development process.
For this reason DAF Trucks has for many years been using simulation techniques in the development and testing of components and complete vehicles, the continually varying operating conditions being simulated under comparable circumstances.
This is, of course, complemented by intensive road testing.

When studying the vehicle's tasks, in fact three basic functions are to be distinguished:
- driving
- supporting
- steering

For research and development of the driving function a computer program called "Route Simulation" has been set up with the emphasis on optimum driveline matching.

Route Simulation, also used by DAF as a sales support tool within TOPEC (the professional transport advice system, including computer programs for vehicle layout, turning circles, driveline performance , route simulation and operating costs) is employed in two ways, depending upon the development phase:
- in the development phase for computing engine and driveline characteristics and matching
- in the testing phase for mechanical endurance tests on the bench.

160 ROUTE SIMULATION AS A PRODUCT DEVELOPMENT TOOL

In both cases the reliability of the results obtained depends to the large extent on the accuracy with which actual operating conditions are simulated.

Much use was made of the Route Simulation model when developing the ATi range, launched in 1985. Since then, the Advanced Turbo Intercooling engines have clearly proved their success in practice.

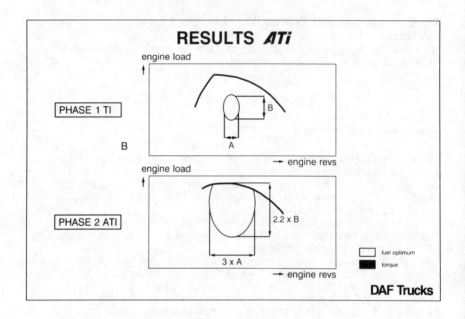

Figure 1. Route Simulation helps specifying the engine characteristics during the development phase.

UTMOST RELIABILITY BEING THE STARTING POINT

To ensure high quality of the Route Simulation model, 95% reliability was set as the minimum standard for it. Equally, the simulation program for the development of engines and complete drivelines, based on varying operating conditions, would have to provide answers to questions such as:
- What are operators' requirements with regard to engine output and torque ?
- Taking the complete vehicle or vehicle combination, what are the load collectives of the different drive components and what is the anticipated lifespan ?
- How do engine characteristics and gearbox and axle ratios influence the engine speed, frequency of gear changing, average vehicle speed and fuel consumption ?

In other words, what is the optimum specification of
the driveline ?
- To what extent can desired product diversification
  be attained with standardised components ?

Collecting all the information needed for all the
different (new) engine and driveline variants in actual
practice would mean having to drive millions of
kilometres in a variety of operating conditions.
Journeys could never be repeated under exactly the same
conditions, and, in view of this time span, they would
be unacceptable from the viewpoint of development
efficiency.
DAF Trucks therefore decided to develop a simulation
technique.

METHOD EVALUATION

Once the "program of requirements" was agreed, time was
spent to build up the philosophy behind the program and
to evaluate possible simulation theories and techniques
on applicability.
In addition to the above mentioned reliability, the
computer model had to meet a.o. the following demands:
- the route data representing the infrastructure should
  not be influenced by the vehicle specification with
  which the data are to be collected;
- it should be easy to compare the detailed calculation
  results with practical experience at any time during
  the calculation process;
- depending on the application, it should be very easy
  to record any specific route by simply driving that.

Evaluation of already existing simulation programs
resulted into further analyses of three basic different
theories:
- the statistical method
- the regression analyses method
- the time based simulation method

Statistical method
The statistical method is based on route representation
defined by a combination of the probability density
function for velocity and for gradients.
Assuming a fixed ratio between velocity and gearbox
ratio (and thus engine speed as well) probability
density functions can be derived for engine-, gear- and
differential load.

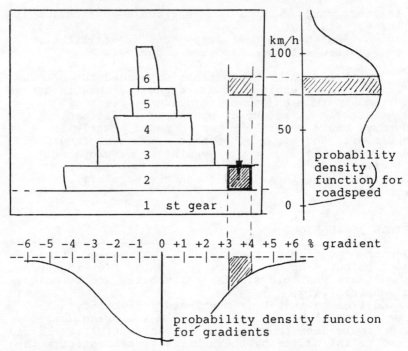

Figure 2. The statistical method is based on routes defined by combined probability density functions.

Regression method
The regression method is based on route representation defined by the probability density function for engine torque against engine speed, also called "load collective".
Generally this function will be derived from torque and engine speed data collected per gear; addition of the data concerned will result into the total engine load, irrespective of it is an increasing gradient or vehicle speed that requires higher engine performance.

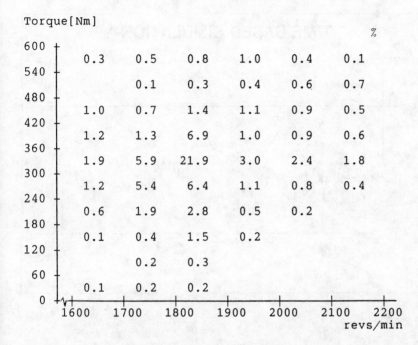

Figure 3. The regression method is based on a route representation defined by the measured probability density function of engine load.

Time based method

The time based simulation is using real road speed and gradient profiles, measured at the same time, related to the actual distance driven. Measurements are not necessarely to be equidistant, however all dynamic variations of road speed and gradient must be covered. The road speed profile represents the real maximum speed to achieve in the particular circumstances, limited by legislation, traffic stops and normal traffic density.

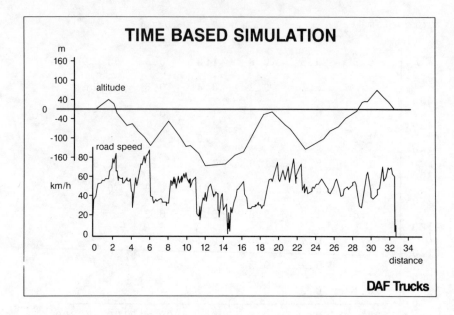

Figure 4. Time based simulation is based on realistic road speed and gradient profile.

After consideration of these methods in view of the criteria mentioned before, DAF opted for the time based method, whereby the behaviour of a vehicle on a particular route is calculated again and again, at very short intervals.

Developing this method is a very complicated and time consuming business, but once it has been created, it provides extremely reliable and comprehensive information.
The intervals at which calculations are made depend upon the number of deliberate or instinctive decisions taken by the driver.

PROGRAM SET-UP

To obtain the basic data for this Route Simulation program, over 120,000 km was covered in different operating conditions on national and international routes. The data for these routes were obtained with data-recorders specially developed for this purpose by DAF Trucks.
The information collected in this way formed the basis for the Route Simulation program.

When developing the computer model great importance was attached to checking of the data obtained while simulating the route.
This is possible because both the route information
- attainable speed
- gradient
and the quantities to be simulated later, the actual
- vehicle speed
- engine speed
- gear engaged
- engine torque
- braking behaviour
- fuel consumption
are measured at the same time, and thus available for analysis of the various routes at intervals.

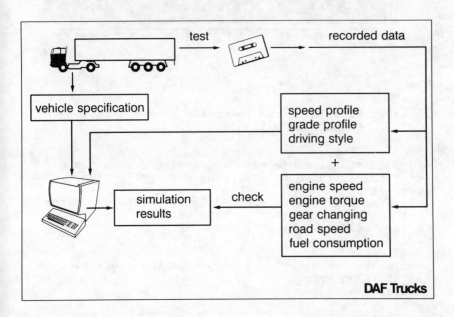

Figure 5. The calculated behaviour of the vehicle and components is continually compared with values measured in practice.

Route data
In order to make the program multipurpose, there have been created two ways to use the various route data that are available in the system:
- A library is available containing many national and international routes through most West European countries, of which one can be chosen, a.o.:
  - London/Rotterdam-Paris-Milano
  - London-Edinburgh
  - Rotterdam-Frankfurt-München-Salzburg-Klagenfurt
  - Madrid-Logrono-Burgos-Madrid
  - Zürich-Basel-Genève-Luzern-Zürich

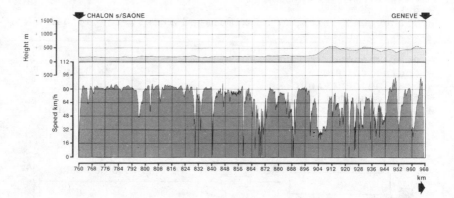

Figure 6. Example of one segment out of the route "London to Milano".

- Any individual route can be compiled from a number of modules, tuned to the transport application concerned.
  These modules are determined by two factors. On the one hand a distinction is made between
  - town roads
  - secondary roads
  - motoways
  whilst on the other, the nature of the route must be established, viz.
  - flat
  - hilly
  - mountainous

Figure 7. Any individual route can be compiled from a number of modules.

Each module has been defined in line with the commonly accepted probability density functions for the appropriate phenomenon, such as gradient depending on flat, hilly or mountainous routes.

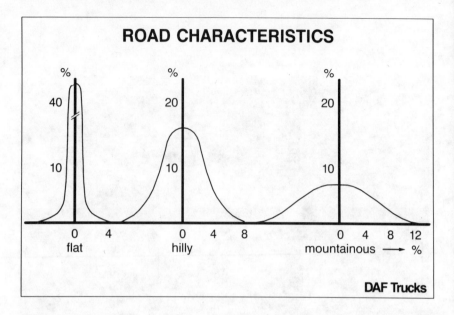

Figure 8. Road characteristics referring to gradients.

Without harming the dynamic speed pattern dictated by the infrastructure, the computer model offers the possibility to indicate a certain vehicle speed limit for each different route segment. Only the maximum speeds will then be modified, being the "target" speed during the simulation procedure.

Driving style
The style of driving obviously has a considerable effect on the final results of the simulation.
For this reason, DAF has done much fundamental research into the factors affecting the style of driving. It was found that at least 21 decision variables together characterise the driving style. These variables relate to the preferred engine speeds in relation to gear changing habits, clutch operation, braking behaviour and the anticipation and assessment of stable traffic situations. Depending on the engine characteristics the decision variables differ from engine to engine.

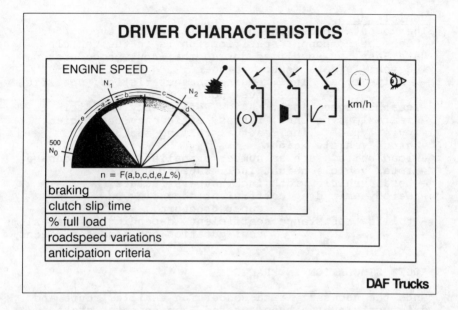

Figure 10. Driving behaviour has a considerable effect on the simulation results and has therefore been integrated in the computer model.

It was also found that the large variety of driving styles arising from the large number of variables mentioned above can be reduced to four basic variants without impairing the reliability of the system as long as all aspects are integrated mathematically. These four driving styles vary from the extremely economical driver who is very fuelconscious to the driver who believes in brisk acceleration and high speeds.

```
DRIVINGSTYLE : fuelconscious --|**|--|-- max.speed
```

Figure 11. The driver chosen here is not the most fuel conscious but still a good one.

Vehicle data

The computer model can handle all vehicle types, for which the component specification is in the Vehicle Data Base, or specified during the input procedure. The vehicle data input includes :
- engine type   incl. engine characteristics, specific fuel consumption
- gearbox type   incl. ratios, efficiency coefficients
- rearaxle type   incl. ratios, efficiency coefficients
- tyre type   incl. actual rolling radius
- gross vehicle weight
- load units, such as number of pallets or net tonnage
- total vehicle height incl. superstructure
- total vehicle width incl. superstructure
- aerodynamic drag coefficient depending on superstructure
- rolling resistance coefficient, depending on the axle configuration and road conditions

ROUTE SIMULATION IN PRACTICE

Once the necessary data concerning vehicle, route and drivingstyle have been entered, the computer starts calculating. At every time interval a calculation is made corresponding the driver's response and the behaviour of the vehicle in the actual circumstances. Acceleration, gear changing, more throttle or less throttle, braking, selection of a particular vehicle speed and anticipation to save fuel are all done with the chosen vehicle on the selected route with the indicated style of driving.
This can only be calculated in detail thanks to the choice for the time based simulation method.

At the end of the journey the results can be presented in various ways :
- Variations in vehicle speed, engine speed and gear engagement during the journey, depending upon the altitude and the desired vehicle speed, can be shown diagrammatically, as shown in figure 12;
- Load collectives for the various components can be shown diagrammatically for the purpose of design criteria and testing instructions, see figure 13;

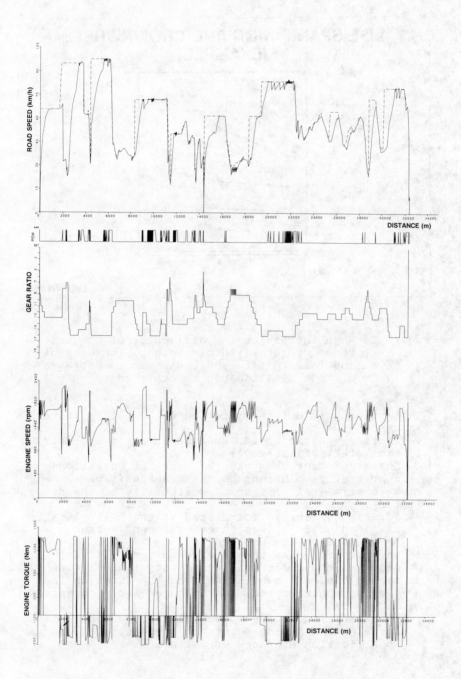

Figure 12. Variations in road speed, gear engaged, engine speed and torque required.

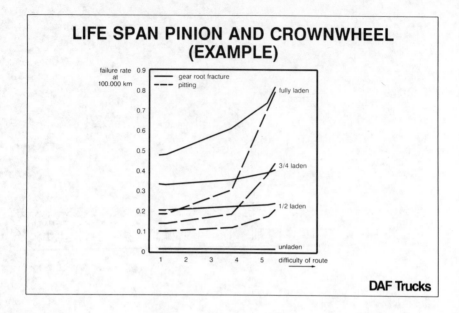

Figure 13. When increasing the difficulty of the test route or total vehicle weight, increase of the failure rate and consequently decrease of test duration can be achieved.

- A tabular printout shows for both the entire route and sections thereof:
  - average vehicle speed, fuel consumption and driving time (ref. efficiency);
  - number of gear changes (ref. ease of operation)
  - number of revolutions of the crankshaft per kilometre (ref. lifespan, suitability of the drive line);
  - efficiency factor, formulated as the multiplication of "load units" times "average speed" versus "average fuel consumption";
  - the percentage of the total time spent in, for example, the different engine output and torque ranges can be shown, as well as the duration in time of engine speed, gear engaged and road speed achieved.
  - the resulting engine load collective which can serve as a starting point for the desired specification of a new driveline.

DAF Trucks/TOPEC — Route simulation

```
VEHICLE DATA
    Engine type                        :DKX - 1160E      [212 kW ECE]
    Gearbox type                       :ZF 16K/S 130     [ 13.68 - 1.00 ]
    Rearaxle type                      :1346 SS 3.31     [  3.31 ]
    Tyre type                          :12 R 22.5        [  0.527]
    Gross vehicle weight       ton  :  38.00
    Load units                      :  25.00
    Total vehicle height       m    :   4.00
    Total vehicle width        m    :   2.50
    Aerodynamic drag coeff.    cw   :   0.85
    Rolling resistance coeff.  kg/t :   6.00
```

ROAD DATA

No.	Tracktype	Distance km	Speed kmh
1	Motorway , flat	220.0	85.0
2	Motorway , hilly	55.0	80.0
3	Motorway , flat	50.0	85.0
4	Interurban, flat	45.0	60.0
5	Interurban, hilly	15.0	60.0
6	City , flat	10.0	45.0
	Total distance :	395.0	

DRIVINGSTYLE: fuelconscious --|**|--|-- max.speed

SIMULATION SUMMARY REPORT

Track no.	Avg speed Km/h	Avg cons. L/100km	Total cons. L	Trip time hh.mm	Number shifts	No. stops	Idle time hh.mm	Revs per km	Eff. factor
1	79.5	34.2	75.2	2.46	59	3	0.02	1041	58.1
2	72.8	35.4	19.5	0.45	36	0	0.00	1111	51.4
3	79.5	34.2	17.1	0.38	13	1	0.00	1041	58.1
4	45.7	28.8	12.9	0.59	65	6	0.03	1537	39.7
5	35.0	52.8	7.9	0.26	47	2	0.01	2240	16.6
6	29.4	41.7	4.2	0.20	69	11	0.02	2265	17.6
Tot	66.9	34.6	136.8	5.54	289	23	0.08	1183	48.3

Note: Figures are provisional

```
Customer name     : COMPAUTO 87
Salesman          : Transport Consultancy Dept.
Vehicle type      : FT 2800 DKXE + 3-AXLE S-TRAILER
Track description : TYPICAL DAYS WORK                    11-DEC-1986 19:41:22
```

Figure 14. Route Simulation summary report.

174 ROUTE SIMULATION AS A PRODUCT DEVELOPMENT TOOL

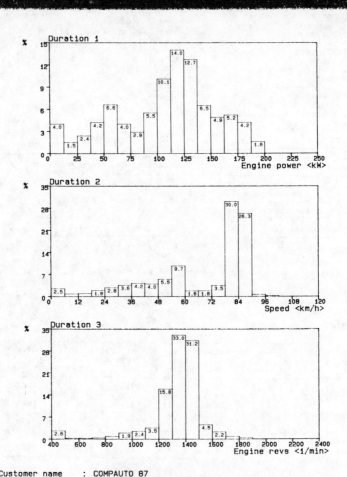

Figure 15. The percentage of the total time spent in,
a) for example the different engine output and torque ranges can be shown as well as the duration in time of road speed or gear engaged.

ROUTE SIMULATION AS A PRODUCT DEVELOPMENT TOOL 175

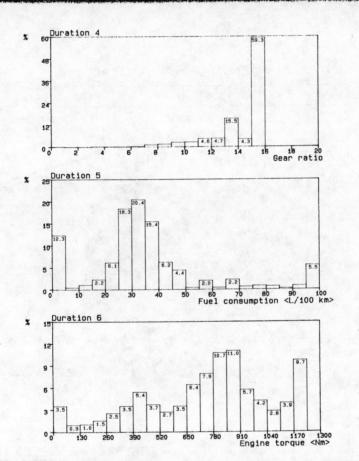

Figure 15. The percentage of the total time spent in,
b) for example the different engine output and torque ranges can be shown as well as the duration in time of road speed or gear engaged.

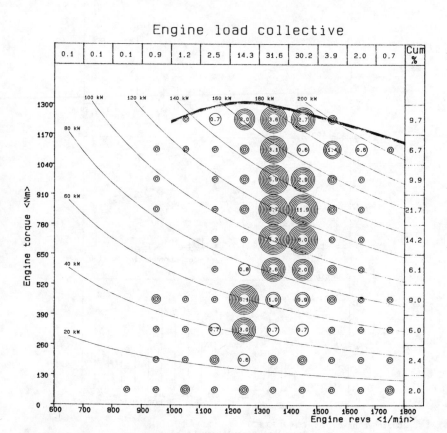

Figure 16 The engine load collective showing the percentage of time spent in a certain combination of engine speed and torque or respective engine power.

ROUTE SIMULATION : FAST AND ACCURATE

As already mentioned, the Route Simulation program forming part of the TOPEC advice system is used to help specifying the optimum vehicle.
For that purpose - as a sales supporting tool - over 3 million kilometres have already been simulated.
The average degree of accuracy has been around 98%. The accuracy for the individual calculations are as shown below.

RELIABILITY EXPERIENCE

- **TOPEC 1984-1985 > 1.500.000 km**   ca. **98%**
- **tests   1980-1985**
  - avg. roadspeed   ca. 99%
  - gear shifts   ca. 95%
  - avg. fuel consumption   ca. 97%
  - load collective   ca. 98%
- **simulation speed by computer**   ca. **7200 km/h**

DAF Trucks

Figure 17. Proven accuracy in practice.

CONCLUSION

The proven accuracy in practice of the individual items calculated, means that DAF's Route Simulation system is a tool which enables development work to be done faster and more accurately than in the conventional manner.
The program is applicable for product development and for specifying the optimum vehicle to customer requirements.

Eindhoven, November 1986

Calculation of Noise Levels in Vehicle Interiors
H. Goldstein
Product Development Systems, Vehicle Design Analysis, Ford of Germany, Postfach 60 40 02, D-5000 Köln 60, West Germany

ABSTRACT

This paper presents FORD's acoustic system model for the calculation of sound pressure levels in the compartment of a passenger car. In this approach the complete vehicle is modelled with chassis, powertrain, body, and interior air space. The structure is represented by vibration modes of the body and a FE-model for the chassis/powertrain subsystem. The structure model and acoustic modes of the air space are assembled to the final acoustic system model. The method allows a wide range of investigations of acoustic phenonena and structural vibrations. Single components of the model can still be added, removed, or modified. For the specific example presented here, a good correlation was found with experimental measurements in the anechoic chamber.

INTRODUCTION

In the past, most of our knowledge about acoustic phenomena in passenger cars has been based on test results. For low frequency acoustics, however, interior noise levels can be calculated by the finite element method using present analysis programs.

The physical process of transmitting sound waves in a passenger car is understood in the following way. When a force, periodic in time, is applied to the vehicle (e.g. engine or wheel forces), structure borne sound waves will travel to the car body, which in consequence radiates noise into the compartment. Air borne sound waves outside the compartment can be neglected here, corresponding to test results in the low frequency regime.

Early finite element models for noise prediction were restricted either to the air space inside the passenger compartment or to the structure of the car. The first type of model was used to calculate and to analyse the cavity resonances in order to identify the locations of the sound pressure maxima. On the other hand, structure models for body, chassis and powertrain were used to investigate noise generating vibrations of the body in response to applied forces. Both models, however, did not accurately predict interior noise.

The calculation of sound pressure levels in the passenger compartment requires a combined model of air space and structure, which mutually interact with each other. This paper describes FORD's acoustic system model for the complete vehicle and illustrates its capabilities for a specific example.

THE SYSTEM MODEL FOR THE STRUCTURE

System models for the structure represent the analytical tool to investigate vibrations excited by external forces. It is assembled from submodels for chassis, powertrain, and car body. Chassis and powertrain are expressed by finite elements representing engine, differential, and all components of the steering column and of the front and the rear suspension. The car body is represented in a different way. The explicit finite element model, which contains up to 30,000 elements (Zimmermann[1]) would be far too large to run it at once. Therefore, structural modeshapes are calculated for the half body model and used as generalized coordinates. Truncating higher frequency modes the model size is significantly reduced.

THE CAVITY MODEL

The modelling process of the air space starts out from the finite element mesh on the interior surface of the car body. The typical mesh size of the cavity model is about 200 mm, which is sufficiently small to describe wavelengths below 1700 mm or frequencies below 200 Hz. Seats are represented by very dense air (FLANIGAN and Borders[2]; Schulze-Schwering[3]) and the effects of panels, that divide the air space into independent regions (package tray and metal sheet at rear seats), are also taken into account.

The cavity model is used here to calculate the acoustic modeshapes in order to generate a modal description of the air space similar to the treatment of the car body. Figure 1 shows a typical example for a cavity mode.

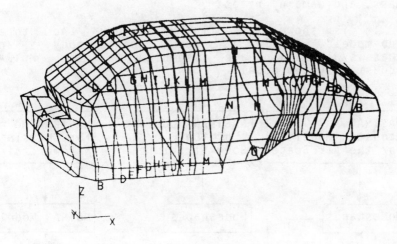

MODE 2   99 Hz

Figure 1. 2nd cavity mode of the specific example. Contour lines on the air space surface denote corresponding sound pressure distribution.

THE ACOUSTIC SYSTEM MODEL

The most important problem in acoustic modelling is to couple the car body and the air space, when both are represented by their modeshapes. The mathematical background for this process has already been outlined in 1980 for the use with MSC/NASTRAN (McNeal, Citerley, and Chargin[4]), which is the finite element analysis program widely used in automotive industry. To date, however, there is no corresponding capability available in NASTRAN.

FORD therefore uses its in-house package MOTRAN, which offers all programs for the coupling process and, in addition, valuable post-processing capabilities for acoustics (Flanigan and Borders[2]). MOTRAN (MOdal and TRansfer function ANalysis) is a software package designed for dynamic analysis of structures, where individual components are expressed either by their modeshapes or by their stiffness, mass, and damping rates.

It is shown in figure 2, how the three sub-models, air space, car body, and chassis/powertrain are assembled to the MOTRAN acoustic system model.

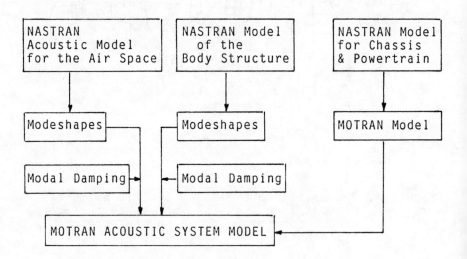

Figure 2. Assembly of Acoustic System Model

MOTRAN utility programs correlate corresponding grid points of air space and body and convert modeshapes calculated with NASTRAN into a MOTRAN component. Each mode of body and air space is damped by a constant fraction of its critical viscous damping. Chassis and powertrain can be modelled in MOTRAN by rigid members connected by springs and dampers. In this approach, however, elasticity is taken into account. The more detailed NASTRAN model, which is part of the structure model described above, is converted into MOTRAN. This part of the assembly process, however, is very time consuming.

Therefore, an alternative method has been developed. It is shown in figure 3. Here car body, chassis and powertrain are assembled in NASTRAN and the modeshapes of the combined NASTRAN model are converted to MOTRAN. The hysteretic damping matrix is calculated with the modeshapes. Structural viscous dampers like shock absorbers are modelled by hand. The final model is then totally expressed by generalized coordinates.

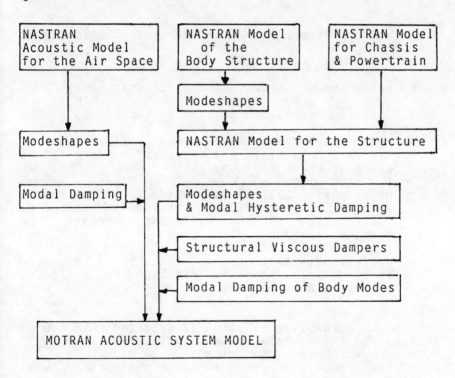

Figure 3. Alternative Assembly of Acoustic System Model

The acoustic system model contains only a few hundred degrees of freedom, so that all the analysis can be performed in the central memory of FORD's Cyber 176 or CRAY X-MP within only a few minutes of CPU-time. The model still allows a wide range of modifications. Any stiffness, mass, and damping rate can be changed, so that single components are added, removed, or replaced. The assembly process has to be repeated only if a new configuration has to be evaluated and if the alternative method has been used.

The most important application of the acoustic system model is to calculate sound pressure levels at selected positions in the compartment as the effect of applied loads varying sinusoidally in time. These results compare with experimental measurements of sound levels in the anechoic chamber. The mathematical method, however, allows to analyse a specific phenomenon in far more detail. At a constant frequency the results are presented in the following form:

o sound pressure levels at selected positions

o structural responses

o responses of structural and acoustic modes

o animated display of structural vibrations

o energy map, which presents the distribution of kinetic, potential, and dissipative energy among all degrees of freedom

o modal participation maps, which split up sound levels or structural responses into the contribution of the different acoustic or structural modes

o design participation maps, which split up sound levels into the contribution of different vibrating body panels

These capabilities are illustrated in more detail for the example in the next section.

Since both, sound levels and structural vibrations, are calculated and since these results can be evaluated by efficient post-processing procedures, the acoustic system model is the currently preferred tool for dynamic analysis.

A SPECIFIC EXAMPLE

In the first application of the method a boom in a rear wheel driven car was investigated. An enhanced sound pressure level was measured at the driver's left ear position at a speed of about 170 km/h or a corresponding first wheel order frequency of 25 Hz. The phenomenon was obviously due to nonuniformities of the rear wheels and the largest sound level was found, when the corresponding forces were in phase. The exact form of the nonuniformity, however, was not known so far.

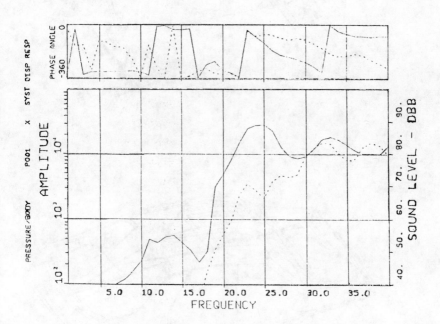

Figure 4. Calculated sound pressure level at driver's left ear for eccentric rear wheels in and out of phase. These results correlate with experimental measurements in the anechoic chamber.

186 CALCULATION OF NOISE LEVELS

The analysis for this example therefore started out from four different load cases with equal amplitudes for both wheels: unbalances, radial runouts, variation of tyre stiffnesses, and eccentric rear wheels. For all cases a similar frequency dependence of the sound pressure level was found. In particular the case of eccentric rear wheels correlated quite well with experimental measurements in the anechoic chamber. Figure 4 presents these results for corresponding in phase and out of phase forces.

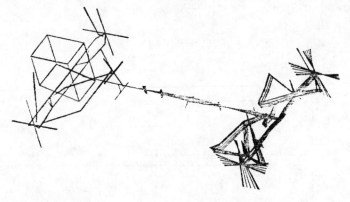

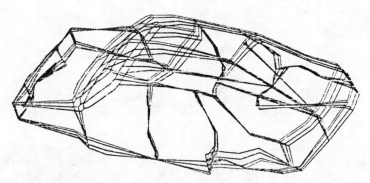

Figure 5. Response of chassis/powertrain and car body.

The next step was to analyse the boom for the in-phase eccentricity in detail. The animated displays of powertrain/chassis and body (figure 5) revealed, that the rear suspension was slightly deformed, whereas the front suspension remained at rest. The motion of the body was dominated by a bending mode (Mode 10). These findings could be confirmed by the energy map and the calculated responses of the symmetric body modes. (Antimetric modes do not couple with the three lowest (symmetric) acoustic modes.) Figure 6 shows resonances of the mode 10, 8 and 3. The acoustic modes were analysed in a similar way (figure 7). Here the volumetric mode turned out to be the leading one. This mode is due to a periodic variation of air space volume, which is for the cavity modes.

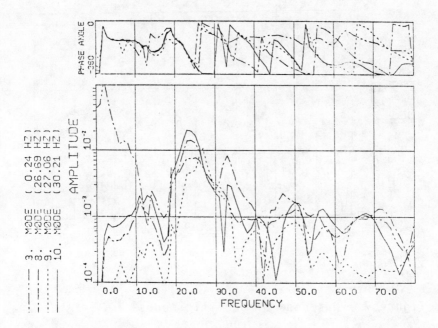

Figure 6. Response of dominating symmetric body modes

188 CALCULATION OF NOISE LEVELS

These results give an overall view of boom phenomenon, but it was still important to know, how acoustic and body modes and the vibration of individual body panels contribute to the sound level at the driver's left ear. This was also calculated using MOTRAN participation programs. The major contributions were due to the volumetric modes for the air space and body mode 8 instead of 10, where mode 8 corresponds to a dash panel and front floor vibration. The most active body area was found on the liftgate.

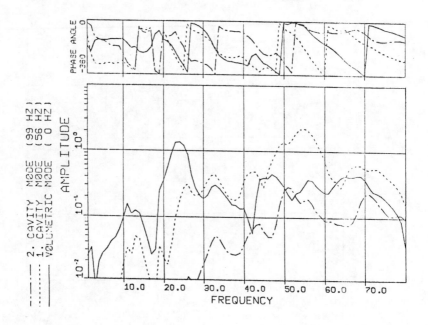

Figure 7. Response of acoustic modes

In order to demonstrate how the method can be applied for noise reduction problems, two different modified configuration were investigated.

The first example concentrated on the chassis, where most of the vibrational energy was found in the rear suspension. Therefore a mass was added to the rear cross member.

In the second example the luggage compartment was stiffened by a massless rigid body, since the vibration of the liftgate turned out to be significant. Figure 8 shows, that both modifications reduced the boom.

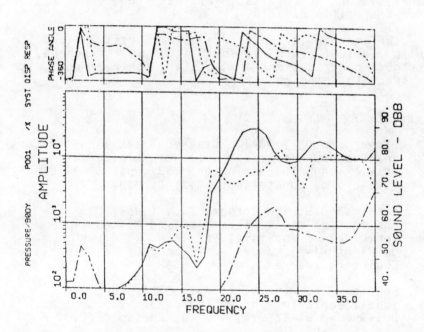

Figure 8. Sound pressure level for modified configurations

SUMMARY AND CONCLUSIONS

FORD's acoustic system model is derived from the structure model and the air space model in a straightforward way by the in-house program MOTRAN. Two different methods can be used for the assembly. The currently used method is less time consuming than the previous one, but the assembly process has to be repeated for a new configuration. Single components, on the other hand, can be still added, removed, or modified.

The model is used to investigate both acoustic phenomena as well as structural vibrations. The method is very efficient, since the model is condensed but still accurate and can be run in the central memory of a mainframe. A number of post-processing capabilities allow a detailed analysis of the results. This has been demonstrated for a specific example.

The method is therefore a very useful tool to gain insight into various acoustic phenomena, to suggest modifications and to study their effect with respect to noise reduction.

REFERENCES

1. Zimmermann D. (1984). Zusammenwirken von Struktur-analyse und Konstruktion in der Karosserie-entwicklung, VDI Berichte Nr.537, pp.289-311, VDI Verlag, Duesseldorf, West Germany.

2. Flanigan D.L. and Borders S.G. (1984). Application of Acoustic Modeling Methods for Vehicle Boom Analysis, SAE Trans., Vol. 93, 840744 (P-144), pp.207-217, Society of Automotive Engineers, Warrendale, PA, USA

3. Schulze-Schwering W. (1983). Berechnung von Schallpegeln im Fahrzeuginnenraum, VDI Berichte Nr.499, pp.91-97, VDI Verlag, Duesseldorf, West Germany.

4. MacNeal R.H. Citerley R. and Chargin M. (1980). A New Method of Analyzing Fluid-Structure Interaction Using MSC/NASTRAN. The MacNeal-Schwendler Corporation, Los Angeles, USA.

SECTION 5 CAD DATA COMMUNICATIONS/CAE SYSTEMS

E.D.I. – Easy Direct Information
J.P. Newton
Geisco Limited, Stockport, U.K.

CONTENTS

1. Introduction

2. What is EDI

3. What is its Impact

4. What are the technical approaches.

5. What products are available

6. Who is using the techniques

7. What are the benefits

8. How do I start

9. What does it cost

10. Summary

1.   INTRODUCTION

The information and perspective presented in this paper have been distilled from the author's experience in developing and promoting Geisco's MOTORNET service for the UK automotive industry since early 1985.

MOTORNET is a clearing house system providing electronic data interchange facilities (EDI) as a part of the Odette initiative in the UK.

Odette UK is a working group of the Society of Motor Manufacturers and Traders (SMMT) and is the national committee of the Odette Europe group which is tackling the issues of creating an electronic trading community across the European automotive industry.

2.   WHAT IS EDI

EDI is the recognised abbreviation for Electronic Data Interchange, also referred to as Electronic Data Communications and Electronic Data Exchange. The term is used to represent the direct linking of one company's computer to another for the exchange of trading information.

Currently the emphasis is to transmit commercial trading documents such as purchase orders, invoices, shipment advices and so on, but already it is being extended to include Computer Aided Design and Manufacture Data (CAD/CAM).

The linking of one computer to another has been going on for several years, using modems and telephone lines the technology is well understood. With the development of the telecommunication provider's packet switched services (PSS's) the basic capability is much more generally available.

The computer industry itself is recognising the need to standardise on communications capabilities, Open System Interfaces (OSI), X25, SNA and similar techniques are now emerging to simplify the problems previously encountered is linking computers.

EDI is the generic term used to encompass all these capabilities and more. Basically it can be considered as a means of controlling computing and communications technology in order that trading partners who wish to exchange information electronically can do so without needing to become experts in the enabling technology.

3.   WHAT IS ITS IMPACT

Consider what currently happens with two trading partners who wish to exchange documents using paper as the media. Generally both have computer systems, if not with different hardware then most probably different applications. To take a purchase order or delivery schedule from customer to supplier involves :-

Customer.

- Prints out the Document from scheduling system.
- Decollates
- Assembles into sets per supplier
- Posts

The Post Office then delivers to the supplier.

Supplier.

- Extracts from the envelope and sorts
- Performs preliminary checks (e.g. batch totalling)
- Transcribes onto data entry proformas.
- Enters into order/production system.

The same transaction exchanged electronically undergoes the following procedure

Customer.

- Extracts from scheduling system.
- Converts into standard format.
- Transmits to clearing house.

The clearing house mails and retains the document until required.

Supplier.

- Receives from clearing house.
- Translates into in-house format
- Inputs to order/production system.

196   E.D.I. - EASY DIRECT INFORMATION

In the electronic cycle there are fewer steps, no need for manual intervention and fewer opportunities for transcription errors.

4.   WHAT ARE THE TECHNICAL APPROACHES

Companies have been exchanging information between computers for many years. Initially data was transferred onto magnetic tapes, these were then sent by courier or post to the trading partner who then loaded them into the in-house system. This was an improvement on using paper, removing the data preparation effort and the need to administer the individual documents. However, as more partners become involved the problems of controlling the number of tapes become significant with the problems of processing each tape individually becoming an increasingly heavy burden.

In addition to using tapes a number of companies, in particular in the automotive industry in the area of dealer systems, have developed services using direct links from one computer to the other. This involves the use of modems, telephone lines and the appropriate software in two computers so that the two machines can 'talk' to each other. This approach has many benefits over the exchange of tapes particularly in the speed of transmission.

It does have complexities of its own: it requires a high level of investment to extend the capability to many users since different computers can require different combinations of modems, transmission techniques and software; it can become difficult to maintain operating schedules to many partners particularly if problems arise or machine failures occur; it can be a highly specialist business to build the facilities to a large population.

In more recent times there has been the emergence of 'clearing houses' which are becoming increasingly recognised as the approach with the greatest mass appeal. Here a network supplier provides a complete store and forward capability for a population of users. The principle is that each user builds and services one link - to the centre. He sends all his data to the centre, it posts it to the various recipients and holds it until it is required. The user can then receive his data, either in total or selectively.

All three approaches are being used today in the market and will continue for the foreseeable future, as the EDI technology emerges and is taken up by the market it is perceived that the clearing house concept will become the dominant option.

5. WHAT PRODUCTS ARE AVAILABLE

EDI involves both computing and communications technology. As the marketplace develops the number of packages and service providers are increasing, these are offering both expertise and cost-effective products and can be classified in three areas:-

5.1 Communications Hardware

Today there is developing a rapid standardisation of communications capabilities. High speed modems operating at 2400 baud and 4800 baud are readily available, off the shelf. For smaller users 1200 baud devices are suitable, generally available, cheap and robust. For large sophisticated operators 9600 baud and kilostream techniques may be appropriate. At the equipment level, subject to speed and compatibility of characteristics, two companies can purchase the appropriate equipment and communicate directly, depending upon the volume and frequency, by leased lines, dial-up services or clearing houses.

5.2 Communications Software

The standard software protocols are also rapidly maturing with the advent of Open Systems Interfacing, X25 and SNA becoming increasingly prevalent. However it will be some time before these techniques are the primary ones in use. For the next one to two years at least 3780 and 2780 emulation will continue to dominate. Here the compatibility between the same emulation on different machines has been less successful and care has to be taken when connecting up trading partners directly to ensure that the two computer operations are compatible.

5.3   Translation and Conversion Software

The need to translate into a standard format and convert from it is key to successful electronic trading around a community. In this area one product currently dominates the market. It is INTERBRIDGE, originally designed and developed by SITPRO, the UK COMPRO and current supplied and supported by a number of companies such as Systems Designers of the UK. This product is table driven and subject to certain rules and constraints it is able to take most computer file layouts and translate into the standard GTDI data structures. There are alternatives such the COPS product produced by Philips in Holland which does an equivalent job. Both products are available on most commercial machines: mainframes, mini and micro-computers.

5.4   Service Providers

A number of computer service companies are developing EDI expertise. I have mentioned Systems Designers with translation and conversion together with an enabling service to modify in-house systems. Geisco, ICL and Istel are active in the UK with clearing house services and can provide expertise and consultancy on resolving the business issues and practices involved with EDI.

In addition the trade associations such as the SMMT in the UK and the various working parties such as the Odette group have gained a pool of experience which is readily available and can be tapped by anyone interested in developing a capability.

6.   WHO IS USING THE TECHNIQUES

Within the UK there have been a number of companies involved in EDI for some time. The automotive industry, with its Odette iniative is in the vanguard of industries promoting and developing EDI technology. Odette UK trialed Geisco's MOTORNET in 1985 resulting in it being endorsed as the UK clearing centre for the automotive industry. A number of companies have been active independently in devising the techniques for themselves but have latterly recognised the benefit of moving towards an industry standard. The active players who have experience of 'private' systems and the general industry initiative are Ford, Austin-Rover, General Motors, Lucas, Perkins Engines, GKN and IMI.

In the UK retail sector ICL is enjoying considerable success with its TRADANET service, again involving many companies who have gained considerable experience building private links and have recognised the benefits of an industry standard public service. Companies who are active here are Sainsbury's, Tesco, Rowntree Macintosh, Marks and Spencers who are all actively encouraging their suppliers to participate.

In addition major British corporations are seriously assessing the benefits of using the technology, such companies as British Steel, British Coal, Rolls Royce and ICI.

The benefits of an international EDI service are being recognised by all the major players and thanks to the efforts of the COMPROS and the EEC commissioners the international trade area is developing very quickly.

Already EDI is being used commercially amongst banks, shipping agents, freight forwarders and carriers for the exchange of such documents as shipping manifests and letters of credit. In this area the potential benefits of fast transmission of data, standard documents and accurate data are immense. Currently it is estimated that 40% of letters of credit are delayed or withheld by banks due to data errors.

In the United States EDI has been established for several years with many major corporations having adopted standards amongst their communities. In Europe the initiative is just beginning to take hold with investigations or activities in process in electronics, pharmaceuticals, chemicals, insurance and distribution sectors.

Whilst today the majority of the effort is concentrated on creating national services there is a rapidly developing momentum to facilitate international trade where the benefits of EDI are immense. This pressure is also resulting in the EEC commissioners themselves looking at EDI techniques as well as the reviewing of customs and excise documentation and the simplication of international trade.

7.   WHAT ARE THE BENEFITS

It is clear from the earlier section on the impact of EDI that the features which make EDI an attractive proposition are the speed and accuracy with which data can be exchanged.

This in turn offers major benefits to manufacturing companies in a number of areas:-

7.1   Clerical Savings
From experience already gained in the UK as part of the Odette initiative in the automotive industry there are significant savings to be gained from the removal of all the clerical effort in administering the paper chain and the keying of data. It has been estimated that for a schedule release the total cost of transmitting the paper document from a manufacturers' computer to a suppliers' is £0.69 per document. Using EDI techniques there is a potential saving of 78% of this cost for national traffic in the UK. For international traffic the costs and potential savings are significantly better.

7.2   Inventory Saving
The same source of experience has also investigated the time delay involved in the paper chain against the electronic exchange and found that it takes 4 days to transmit the paper document and less than 1 day by electronic means, again looking at goods shipments around the UK only.

Therefore an immediate improvement in stock holding of 3 days at no cost to service level.

7.3   Significance of Just-In-Time
These savings of 78% in administration overhead and over 3 days in transmission time are recognised as only the beginning of the story. The opportunity to integrate manufacturers and suppliers production systems, rationalise the number and complexity of documents exchanged and improve the quality and accuracy of data available are seen to be the real benefits of introducing these techniques. It is generally acknowledged amongst those companies involved that EDI is an essential and significant step in

developing the Just-in-time philosophy of manufacturing across the European Automotive Industry.

8.   HOW DO I START

The issues which need to be addressed before a company can gain the benefits of EDI are as follows:-

1. With whom do I wish to trade: establish the trading partners (customers and suppliers) with whom you wish to exchange documents electronically. Discuss with them the type, structure and content of each document in terms of the data carried.

2. Agree the standards to be adopted: Increasing in Europe a document standard called GTDI is used but more recently a United Nations committee on EDI has proposed a worldwide syntax referred to as the JEDI standard. More details of this are available through the national COMPRO organisations such as SITPRO in the UK.

3. Set a target and timetable for all the players: The introduction of EDI requires the maximum co-operation of everyone involved. Therefore it is important to create an environment where all the participants understand the issues, are committed to the approach and share the goals.

4. Organise the development: Obtain the necessary hardware, communications facilities and software to facilitate the service. Here it may be necessary to involve a number of service providers to enable progress to be made to introduce such facilities as: a clearing house; conversion and translation facilities into and out of the standards; support on the implementation of the technology required to create an integrated operation. A number of companies are already active in this area, advice is available from a number of the trade associations connected with the various initiatives so that once the

previous stages have been completed this becomes a straight forward step.

5. Conduct trials: Before a fully live operation is achieved an appropriate time of at least one month should be allowed to test the implementation. During this period the emphasis is on the organisational impact of using EDI and the testing of the new procedures and controls required. Today, the technology is already proven and working.

6. Live operation: It must be recognised that the introduction of EDI will require an extended period as each community will expand from a small core of 'drivers'.

7. Joining Existing Communities: If a company wishes to join an existing service then many of the above steps will have been previously completed by the participants of that service. The task for a new user is then to understand the standards employed (step 2) and organise his own operation (step 4).

9. WHAT DOES IT COST

Whether a company is beginning an initiative on behalf of a community or joining an existing service the direct investment costs are similar. If a company is an existing computer user the cost of modems, communications software, translation and conversion software and changes required to in-house systems is of the order of £5000 - £10,000 sterling. For a micro-computer the equivalent cost is £2000 - £4000 sterling. Obviously this will vary from one company to another and will require some investigation of existing capability and capacity. The service providers are geared up to offer this advice and consultancy. Using clearing houses direct costs are pennies per transaction, using other methods investment costs can be significantly higher subject to the method chosen, but operating costs can appear very low.

There is additional cost of resources and time to adjust the internal procedures and controls in line with the impact of electronic trading. This will obviously depend upon current situations and practices but it is fair to say that once companies have completed the transition stage the costs incurred in changing the administrative system are very quickly recovered in the new operating environment. Again the various service providers plus the trade associations can give advice and direction in this area.

10.   SUMMARY

Electronic data Interchange (EDI) is the means by which trading partners can exchange data between computers electronically. Currently the emphasis is on commercial transactions (for example invoices, credit notes, shipment advices) but the areas of CAD/CAM are being investigated. The objective is to remove paper as the primary media for exchanging information and to simplify the technology to make electronic transmission a simple and cheap facility for everyone.
Using paper the transmission of documents on paper is a long, clerically intensive exercise with many opportunities for translation errors and delays. Using electronic links the exchange of data between two computers is fast, less error prone and requires no manual handling of data.

EDI can take three forms: the production and exchange of tapes between companies; the direct linking of one computer to another using modems and telephone lines; the use of clearing house to act as a 'store and forward' service. Whilst all three options are in use today and will be for the foreseeable future it is increasing recognised that the clearing house approach has the greatest mass appeal.

The equipment required is readily available now: modems for a whole range of uses; leased lines; dial up connections and Packet Switched Services. As well as the hardware, communications software is required to support the various emulators and protocols prevalent in the industry. From the current 3780 and 2780 style which will continue to dominate for the next one to two years to the fast developing OSI capabilities of X25 and SNA. Software packages to translate into the standards and convert from them are available: INTERBRIDGE developed by SITPRO and supplied by such companies as System Designers in the UK and COPS developed by Philips in Holland. Clearing House Services together with expertise are being offered by Istel, ICL and Geisco.

The benefits of using EDI are substantial; a saving on clerical administration of 78% per document and an improvement on the speed of transmission from 4 days to less than 1 day. The consequential effects in terms of administrative and inventory savings whilst being important are not the major achievement, the benefits from the integration of manufacturers and suppliers system, of rationisation of documents and improvements in data accuracy make EDI a very signficant step in the development of Just-In-Time operations.

In the UK several communities are already active with EDI techniques: in the automotive industry, Ford, Austin Rover, General Motors, Peugeot/Talbot and their major suppliers; in the retail sector, Sainsbury's, Tesco, Rowntree Macintosh, Marks and Spencer with their trading partners. In the United States EDI has been established for many years with major corporations promoting their own services. More recently there has been increased pressure towards standardisation and inter-link separate communities. In Europe and internationally the benefits are now being recognised particularly in import/export documentation such as bills of lading, letters of credit and shipment manifests. This pressure has resulted in the EEC commissioners looking seriously at EDI as well as simplified customs and excise documentation.

Companies who wish to use EDI fall into two categories: those who need to establish the facility across their trading community and those who simply need to join an existing service.

For those who need to build a community there are several steps involved in creating an operational service which require close co-operation amongst the drivers and the involvement of a small number of service providers to create the necessary facilities. To join an existing service a company simply needs to adopt the appropriate standards, develop the computing and communications capability and adapt its procedures and controls.

The investment costs associated with introducing EDI are as little as £2000 - £5000 sterling for a micro-computer users to £5000 - £10,000 sterling for a mini or mainframe computer. There are hidden costs in adopting new control and procedures but experience to-date has been that these are very quickly recovered from a more streamlined and effective operation.

Electronic Data Interchange is creating an easy path towards cheaper, faster and accurate data exchange. The technology is available today. It needs more companies to join in, gain the experience and encourage the development. The concept is a major initiative in creating more cost effective trading relationships, this will affect everyone who trades and the earlier users will gain major competitive advantages.

CAD Data Communication
M.J. Newton, K.S. Hurst
*Department of Mechanical Engineering, Sunderland Polytechnic,
Sunderland SR1 3SD, England*

SYNOPSIS

In order to optimise the benefits of CAE in the broadest
sense for both motor manufacturer and component supplier, it
is essential that information captured during the design and
draughting stages be fully transferable between companies.
The modern motor vehicle is complex with components designed
and manufactured by a network of suppliers or subcontractors.
It is therefore a logical step to transfer CAD data in both
directions between motor manufacturer and component supplier.

There are various avenues which, theoretically at least,
are open to consideration in achieving this aim. The component supplier could use a bureau; lease a system; purchase
the same system as the co-ordinating company; purchase a
different system incorporating interfaces with the co-ordinating company's system, or gain access to the main manufacturing data base. The purchase of a CAD facility, by a
supplier, would allow development of in-house capabilities
and would undoubtedly increase the efficiency of the supplier,
but would require expenditure of comparatively large capital
funds which a small company could find difficult to justify.

The immense potential benefits of CAD data communication between motor manufacturer and supplier are identified
and problems of data exchange are reviewed. The paper
identifies the possible means by which the whole industry
may overcome the current problems of data communication.

INTRODUCTION

The introduction of CAE into a wide range of engineering industries has taken place over recent years. Perhaps the greatest expenditure to date has occurred in the area of CAD with numerous companies having completed the transition from conventional drawing offices to modern electronic offices. Initially these CAD systems were largely utilised for 2D draughting but with the rapid development in both software and hardware many CAD systems now support 3D solid and surface modelling, design analysis and manufacturing software.

The extensive use of CAD within any organisation generally involves access to central data bases and often the transfer of data between computer systems. This latter situation frequently leads to data communication problems even when the data is transferred between so called compatible systems. This problem is further exacerbated when data has to be transferred between dissimilar systems or from one company to another where a dissimilar CAD system may be in use.

The automotive industry is almost unique in that it comprises a number of manufacturers who purchase components from a large number of suppliers, all of whom are concerned with the mass production of components. The conceptual design is generally undertaken by the motor manufacturers and components designed by the suppliers within this inital framework.

Most motor manufacturers are now well entrenched in the area of CAD, with extensive computing capability. The logical step is to maximise on this by transferring CAD data direct to the component suppliers who can then utilise this information to design their component(s), returning data in a similar manner. This, of course, requires the suppliers to equip, or gain access to, a CAD system. This may pose a significant difficulty for some suppliers faced with the apparent choice of either purchasing a relatively expensive CAD system or losing the contract.

This paper aims to clarify the potential benefits of direct data transfer, outlines the various options open to the supplier companies and makes suggestions as to the way forward towards the optimum solution for both interested parties.

THE BENEFITS OF DIRECT CAD DATA COMMUNICATION

During the initial evolutionary design period a great deal of information is generated. One of the major advantages of any CAD system is that once data is captured it need never be recreated and modifications or further development can be carried out rapidly. It is desirable therefore to allow supplier companies access to the design data base. This eliminates the need for redrawing by both the supplier and

the manufacturer. As a direct result of this process the risks of
misinterpretation of design information are minimised and the
speed of response of the supplier, with both technical data and
quotations, can be greatly improved, particularly where the information is complex in nature. In addition the employment of
a CAD system enables drawings to be produced to the same high
standard and quality.

A further significant benefit of direct data communication
is that of increased security since the need to distribute hard
copies of drawings is reduced. This is particularly important
in the development stages of a project.

It should be recognised by all parties however, that only
design data held within the system can be transferred. There
may still be the need to send supplementary information in the
conventional manner.

These benefits are perhaps obvious but problems can, and
frequently do, arise when CAD data is to be transferred to
other companies or suppliers. The lines of communication can
be complicated, suppliers often having to deal with more than
one department within a large manufacturer. If the supplier
has to deal with more than one manufacturer, each of which may
be using a different CAD system, then the problem becomes
compounded.

The state of the art in CAD is continually advancing and
if the benefits mentioned above are to be realised then the
suppliers must inevitably possess a capability in the area of
CAD (and possibly CAM). Some suppliers may already have this
capability and some may even have compatible systems, but what
of the rest?

POSSIBLE OPTIONS FOR CAD DATA TRANSFER

The following are some of the options open to the suppliers:

1.   Purchase an identical CAD system to the motor manufacturer.
2.   Purchase a different system to the motor manufacturer.
3.   Use a bureau service.
4.   Direct on line access to a central data base.
5.   Lease a system.

How can a supplier company make a sensible choice from
these options? It is essential that at least a basic understanding of CAD (and its jargonistic terms) is gained before
entering into discussions with any interested parties. Some
companies may already have experienced personnel in CAD, some
may find it beneficial to be educated in the subject, whilst
others may wish to be guided by reputable consultants. The
latter would be particularly beneficial if the supplier company were being pressurised into purchasing a system at

relatively short notice.

The initial requirement for a CAD system may either come from the supplier or the motor manufacturer. Whichever the case, it is sensible to choose one of the above options with the requirement of data communication firmly in mind. Each of the options are now examined with consideration of their relative advantages and disadvantages.

THE PURCHASE OF AN IDENTICAL SYSTEM

In the main this option may only be viable for large suppliers and/or subsidiaries of the parent company since the cost of purchasing and maintaining a powerful 3D system would be difficult to justify. Also if the company supplies more than one manufacturer there is no guarantee that the motor manufacturers will have identical CAD systems. Even when running the same software on different hardware data transfer problems can be encountered.

A further problem which could be encountered by the motor manufacturers is that if the only realistic option for the suppliers involves large expense, effectively restricting the number of companies capable of responding, then the policy of dual sourcing may be rendered impossible.

One way of making it possible for supplier companies to purchase the same relatively expensive system(s) as the manufacturer is for all parties to unite, perhaps under the umbrella of the Society of Motor Manufacturers & Traders (SMMT) and to negotiate the purchase of the same system from one vendor. This could bring with it the cost savings usually encountered when bulk buying and may bring even the most expensive systems within the range of the smallest suppliers. All of this does assume that every motor manufacturer uses the same CAD software.

THE PURCHASE OF A DIFFERENT SYSTEM

It is essential for all interested parties to discuss the type, volume and format of data to be transferred to enable an initial specification to be drawn up and then sent to CAD vendors.

There are many CAD packages which are now highly developed and well proven. The main problem is incompatability of data format. The claims of vendors must be checked by undertaking rigorous benchmarking. Failure to do this in the past has caused significant problems for many companies and has led to the Department of Trade and Industry (UK) producing videos and text on how to undertake successful benchmarking. Only after comprehensive benchmarking should a short list of suitable vendors be compiled.

Assuming data transfer between dissimilar systems to be a practical proposition, either direct or via a graphics standard such as IGES, this option is possibly the most attractive alternative for most suppliers and the smaller suppliers in particular. The attraction is not solely one of cost. A neutral format, such as IGES, with its pre and post processor, also has the advantage of freeing the user from dependence on obsolete processors since data/drawings can be archived in the neutral format and transferred to any new system. The standardisation on IGES has been fraught with difficulties and many vendors have supported or developed an alternative system.

An excellent review of the current status of IGES with particular reference to the motor trade was presented by W. Owen (1). One statement in this paper attributed to John Owen of the CADCAM Data Exchange Technical Centre was that he was "not aware of any full implementation of IGES 3.0". It is claimed that IGES 4.0 will address the problem of solid model data transfer although a change of acronym to Pdes (Product definition exchange standard) is forecast.

A truly international standard, Step (Standard for exchange of data) based on Vda (the German standard), Set (Standard d'échange et de transfer) and Pdes is now under discussion by the International Standards Organisation (ref.1). However, even with all of these grandiose plans the Society of Motor Manufacturers and Traders became so disillusioned that they threatened an embargo on CAD vendor's products (ref.1).

Some Independent Design companies in the motor industry have spent large sums on computer equipment and have tackled the problem of data communication in a slightly different manner. Instead of using neutral standards they are developing their own interfaces when and where they are needed. Indeed P. Wylie (2) in his paper states that International Automotive Design (IAD) have "a full-time programmer devoted to developing interfaces to allow communication between different types of computers". IAD have also found 3D communication to be easier than 2D because of the lack of dimensions, etc. The possibility of the suppliers employing their own programmer is only viable in the case of the large suppliers. Numerous vendors have, however, developed translators which will interface with many of the popular CAD packages and this is becoming a popular solution to the data exchange problem as more systems are implemented.

It is interesting to compare the Aircraft Industry with the Motor Trade. According to Hitch (3) British Aerospace (BAe) and Aerospatiale (France) have many diverse CAE systems and in 1983 began to develop a common communications format for data exchange. BAe have chosen Set in preference to IGES when working in Europe.

USE A BUREAU SERVICE

This is probably the most realistic solution to the problem for the small supplier, particularly if the manufacturer insists on communication in 3D. It may even be the best short term solution for large suppliers until a firm decision is made by the manufacturers as to the level at which communication will take place.

The disadvantages of using a bureau include possible security problems, ongoing costs, slow response times and a loss of all the potential benefits which an in-house system would bring.

DIRECT ON LINE ACCESS TO CENTRAL DATA BASE

Within large engineering organisations different departments may possess local computing capability. Frequently these local facilities are networked enabling users in different departments to access data contained within the network. Since all departments in the local area network are directly connected with each other, there is no necessity to transfer data by tape.

One possible solution in overcoming the data transfer problem to and from suppliers is to extend a network to embrace supplier companies. At first sight this may seem totally unworkable since it could in reality involve hundreds of suppliers being connected to a central facility. There would also be the need for a stringent security system to be in operation.

The benefits are largely to the supplier in that it would dispense with the need of each to purchase a free standing CAD system. A CAD workstation, peripheral devices and modem is all that may be required. The central computing facility would be the hub of the network and the costs could be distributed between the suppliers.

This option can be likened to a number of users having access to a bureau service for which they pay for logged CPU time. The motor manufacturers, some of whom own bureaux, could organise this facility or it could be developed by the suppliers in conjunction with a central coordinating body.

Of all the options presented this is perhaps the most ambitious and whilst it has great potential for the industry, could fail unless correctly planned and fully supported by all concerned from the outset.

LEASING A SYSTEM

This is not always possible as some vendors do not offer the option of leasing. However, where they do the long term costs of leasing are inevitably higher than those incurred in the purchase and subsequent maintenance of a CAD system. This option does however allow a more flexible approach to be taken by all of the interested parties. As previously mentioned the selection and purchase of a system must be carefully undertaken and even then, data transfer problems may be encountered. Leasing allows a thorough evaluation of a CAD system, demonstrates the willingness of the supplier to meet the manufacturer's requirements and above all delays the purchase of a system until the present problems of data transfer, previously outlined, have been resolved.

Leasing also has the advantage over use of a bureau service in that during the period of the lease, in-house experience in CAD is being developed. This would enable a more subjective approach to be taken when the decision to purchase a system is finally made.

DECIDING ON THE OPTION

The final choice will depend on a number of factors and throughout the industry a mix of the options previously discussed may result.

In the first instance the interested parties must discuss their requirements either on a one to one basis or in a forum. Work is already progressing in this vein through the ODETTE Engineering Sub-Group. When systems are finally being purchased or other costs borne by the supplier, it is the individual supplier who must be convinced that the best solution has been chosen. It is important that the supplier companies fully evaluate their own requirements and not just the needs of their customers. It is not only the initial outlay that must be considered, the recurring cost of maintenance (typically 10 to 20% of the purchase price per annum) must be met.

The choice is not an easy one when faced with the problems of data communication, number and variety of systems on the market and internal cost justification. One thing is certain, however, and that is that if suppliers are to survive, most, if not all, will have to possess or gain access to a CAD system in the not too distant future. If a supplier does not possess CAD expertise, and since a little knowledge is dangerous, it may be prudent for the supplier to seek support from reputable independent advisers.

CONCLUSIONS

Computer methods are being used at a very early stage in the design and development of new vehicles. It has been traditional that a full size "clay" be made during the development of a new body and there is still no substitute for the physical model. However, the way that this is produced is changing. Now the stylist can work on a computer based surface modeller and the data generated used to drive a DNC machine to create the physical model. Once the model is accepted, the data base can be accessed by jig and tool designers, planners and estimators and of course the designers for solid modelling and finite element analysis.

It is tempting for the motor manufacturers, particularly in the development stages, to want to send information to potential suppliers in this 3D form and not to produce 2D detail drawings. The current cost of 3D systems and the requirement for 3D data transfer, would effectively limit the supplier to the same system as the manufacturer or the use of a bureau. It is therefore possible that in order to optimise the overall benefits of data transfer that the manufacturers should accept the little extra work required in order that communication may take place in 2D, where 3D is not essential.

Even with the potential advantages of bulk buying (as suggested earlier) supplier companies will find cost justification of a CAD system based solely on data transfer with motor manufacturers virtually impossible. This approach would also be extremely short sighted since the potential internal benefits could far outweigh those from external data transfer. Primrose, Creamer and Leonard (4) listed the internal benefits briefly as: reduce DO staff and clerical labour elimination of model making, faster quotations, new products introduced more quickly, component standardisation, design optimisation and better estimating. Whilst some of these claims are unrealistic they also ignore the enormous benefits of linking through to manufacture (CNC and DNC) and inspection/test (CADMAT).

As a final recommendation it is essential for the manufacturers and suppliers to get together at an early stage in order to optimise the enormous potential benefits of electronic data transfer to their mutual advantage.

REFERENCES

1. Owen, W. The Data Format Puzzle, CADCAM International, Sept. 1986.

2. Wylie, P. Modelling the Shape of the Motor Industry, The Engineer 24/31 July 1986.

3. Hitch, H.P.Y. CADCAM in the Aircraft Industry, Proc. I.Mech.E., Vol.200, No.B2.

4. Primrose, P.L., Creamer, G.D. and Leonard, R. Identifying the Company Wide Benefits of CAD Within the Structure of a Comprehensive Investment Program. CAD Vol.17, No.1 (Jan/Feb 1985).

Application of CAD to Autobody Design in a Commercial Multi-Client Environment
D.J. Clark
IAD Computer Systems Limited, Dominion Way, Worthing, England BN14 8LU

Introduction

As the product of the automotive industry has increased in complexity, so the organisational complexity of production has increased. Part of that organisation has been the generation and development of client-supplier relationships into an intricate web of interconnections. While communications along the web used a common medium, that is hard copy, data exchange was relatively freely carried out. The advent of Computer Aided Design (CAD) and its adoption in the automotive sector of the engineering industry has led to new strains being exerted on the established client-supplier links and a change in the interdependence of each party through that link. In the past clients were relatively free to work with the full range of suppliers and vice versa. This choice is now being restricted to those parties working with a compatible data medium.

International Automotive Design (IAD) are consultant design and tooling engineers based in the U.K. and have been active internationally in this field during the transition period introduced above.

The experience gained by IAD has extended across many fields of the automotive industry, but for this paper I shall restrict myself to the field of auto-body design where my own experience has been gained.

In recent years it has been the declared policy of Austin Rover Group Limited to pursue the implementation of a Computer Integrated Technology strategy and in the view of the company directors there has been considerable progress toward achieving this aim. The majority of the remaining established car manufacturing companies have adopted a similar policy although implementation has followed differing paths and reached differing levels of completion. Naturally with the large capital base of many of these companies a high level of investment in expensive new CAD technology has been possible. It is this level of investment which has proved difficult for the myriad of small suppliers. Thus the intricate web of relationships existing in the automotive industry is changing so that the declared policy of the majors can be fully implemented.

BROAD ASPECTS OF APPLICATION

The problems of data exchange may appear to be sufficient deterrent to the installation of CAD systems, however, the potential benefits in engineering terms alone far outweigh the short term commercial problems. With hard-copy data formats the information conveyed is restricted to discrete sections in two dimensions which can be assembled into a three dimensional wire frame. Thus the surface area between the frames is open to interpretation by the panel manufacturers. With a full three dimensional surface modelling CAD package it is possible to specify the entire surface at the design stage and with Computer Numerical Control (CNC) post processing to produce a part completely within design specifications.

CAD surface modelling techniques have developed to high levels through applications across the engineering spectrum, specifically those which deal in surfaces of high complexity such as automotive, aeronautic and marine. The quality of surface definition is paramount in application competence and the achievement of this aim has been cited as one of the major factors in the array of differing mathematical bases to alternative surface modellers, each purporting to have strengths over its competitors in certain specific applications. Many engineers will be familiar with the terms Coons patch, Gordons patch and Bezier surface if not with the mathematical strengths and weakness of each form.

Ford have invested a large amount of development time into their own C.A.D. system. Known as the Product Design Graphics System (PDGS) it has a variety of capabilities including a 3D surface modeller and a related numerical control module for generation of CNC cutter paths. Based on Cincy parabolas acting as bounding functions for each surface the basic form can be modified interactively to suit design requirements. The resulting surface representation can excellently describe the body panels required, but is stored in a form totally incompatible with the majority of other surface modellers. This does not imply that the remaining surface modellers are compatible, quite the reverse is true and I shall deal with this problem in greater detail later in the paper.

The majority of established surface modelling packages are available to be installed on a specific hardware type different from their competitors. Thus if a supplier requires a different CAD system to be compatible with their clients it does not solely necessitate new software, but also additional or different hardware. A range of surface modellers with their preferred hardware type is given in TABLE 1.

FACTORS IN THE APPLICATION OF INDIVIDUAL SYSTEMS TO AUTOBODY DESIGN

In the application of individual systems, each has its own advantages and drawbacks, however, the approach should be common. The mathematics in general should be invisible, as body engineers are best occupied overcoming engineering problems.

SURFACE MODELLER	PREFERRED HARDWARE
CATIA	IBM
CADAM	IBM
STRIM	IBM
SYSTRID	IBM
PDGS	PRIME/LUNDY
CADDS 4X	COMPUTERVISION
INTERGRAPH	INTERGRAPH
SWANS	VARIOUS (PRIME, DEC, ETC)

TABLE 1 : SURFACE MODELLING SYSTEMS AND PREFERRED HARDWARE TYPES (GUIDE ONLY)

Certain systems run from a user command string which is closely related to the mathematic control parameters. Many of these do not directly relate to engineering parameters and so the operators require additional skills to complete even the most basic of tasks. Clearly the most successful surface modellers are those which allow the body engineer to create parts in a manner which is easily understandable to him or her, without the need for excessive training. Apart from the apparent user friendliness a system should included an array of functions which relate directly to engineering/draughting concepts. Functions such as fillet generators are particularly useful as an aid to rapid part production.

Certain mathematical formulations can hinder progress by actually generating irregularities within the surface despite smooth boundary lines. The Bezier surface in Figure 1 can be seen to contain ripples indicated by non-smooth surface lines. These surface imperfections originate in the complex mathematical formulation and not in any engineering data input. Such blemishes are generally unacceptable and can require much skill and effort to remove. The skills required include a fair degree of understanding of the numerical form in which the surfaces are stored. Recent developments in mathematics have led to the development of a process known as reparameterisation which aids the designer in overcoming problems such as these.

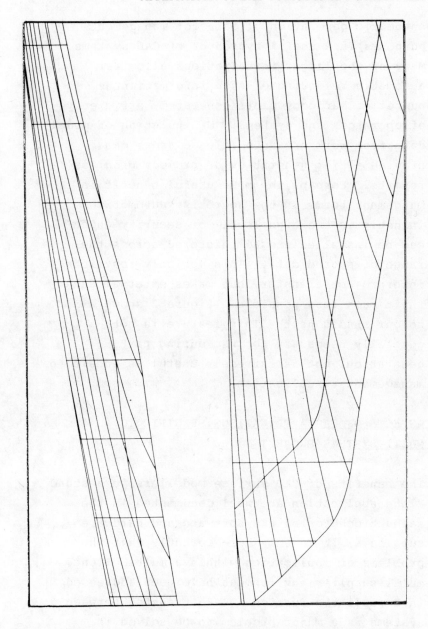

FIGURE 1 : EXAMPLE OF MATHEMATICS GENERATED SURFACE IMPERFECTION

Automatic generation of data for management purposes is a useful byproduct of CAD systems. Many of the larger installations allow for automatic recording of such information as number of hardcopy plots generated, efficiency of operators and systems. Manipulation of model data from work areas to release areas can provide easier generation of project progress reports. Perhaps the most useful benefit in mangement terms in the secretive automotive industry is the high level of security which can be installed and administered across the range of stored data. This not only restricts interproject transfer, but makes external data theft much less possible. Figure 2 shows a colour shaded part. This feature is not generally necessary for use during part generation, but is extremely useful as an aid to management visualisation.

ASPECTS OF CAD APPLICATIONS SPECIFIC TO MULTI-SYSTEMS PROJECTS

The benefits of CAD surface modellers in a stand alone application are well documented in the sales brochures of the software and hardware suppliers; it is only more recently that the problems of application within a multi-client, multi-supplier environment have been addressed in a realistic fashion. Data transfer between systems is a major problem to be solved in general terms, if not in detail at the earliest stages of a project. To obtain at least the minimum benefit from CAD operation, then part geometric data should be transferable.

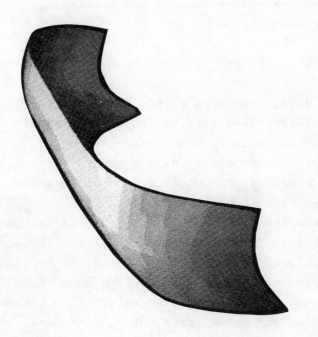

FIGURE 2 : EXAMPLE OF SHADED
SURFACE TO AID VISUALISATION
BY NON TECHNICAL STAFF

Of lesser importance in the body engineering sector are drawing entities such as text and notes. There are several physical methods by which the data can be transferred, each having strengths and weaknesses for particular applications.

Transfer of Data via Hard Copy

New developments in raster to vector software technology have lead to the automation of hardcopy input to CAD databases. Several companies now offer fully packaged systems. It is generally acknowledged that these systems can never hope to match the accuracy of the CAD system itself as the hardcopy medium is open to distortion and degradation if stored for long periods of time. If the only source of the information is in hardcopy form then these systems are an attractive alternative to manual scaling and manual data entry. The same problems of media degradation affect manual data entry. It is possible with some systems to calibrate the raster information by applying creep or shrink scale factors such that the resultant CAD vector information is more true to scale than the hardcopy draught. This method could be used to transfer data from CAD system to CAD system, however it is restricted to discrete two dimension/three dimension wire frame and thus is not applicable to surface transfer.

Data Transfer by Magnetic Media

Data transfer by magnetic tape has been used extensively to link computer sites in an off line manner. Where the sites run similar software at similar release level on similar hardware this method is relatively rapid and requires little preparation. Data can be loaded and unloaded using functions similar to the backup and restore utilities provided with all systems. However, the majority of sites which operate in a multi-client/supplier environment must liaise with installations of a different type. As a result an agreed data format must be established whereby each system can access the information in the required manner. The ultimate objective of such a system must be total information transfer, however, in the majority of cases a subset of the whole may be acceptable. The format for this data transfer has been the subject for a large amount of discussion in most centres of engineering worldwide. Many centres have developed their own format sufficient for their own needs, lately there has been an initiative towards the development of a single global standard based on the strengths and weakness of the present formats. The range of applications to which this standard is to be introduced has added to its potential complexity and also to its projected leadtime. With its roots in early interfacing work carried out by several separate bodies around the world the projected family tree of the "Universal Standard" ISO STEP can be seen in Figure (3).

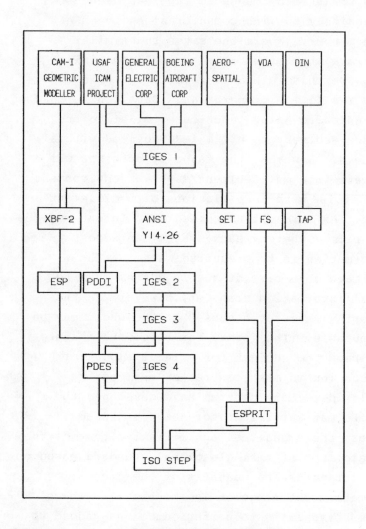

FIGURE 3 : INTERRELATIONSHIP OF DATA INTERCHANGE STANDARDS

The figure shows the standard formats which have been developed with a view to being industry standards. There are in addition such formats as Ford Standard Tape and Austin Rover Neutral file which are typical of company internal standards and further add to the complexity of this field. Recently the initiative in development of data standards has been followed up by large user groups exerting presssure on CAD suppliers to provide interface routines to international standard formats such as IGES 3.0 as part of their products. The problem of data format input to most machines in solvable through software development and installation, however if the tape units associated with these systems are rigidly set up to write for example at differing densities or using special labels then transfer may have to be carried out by other means.

Transfer by Telecommunications Media

Electronic data transfer via communications lines, whether including satellite, dedicated lines or a public data link is an obvious method of avoiding the above problem. It remains subject to the establishment of an agreed geometric data format, but is generally more rapid with an associated extra cost penalty. Whereas with magnetic tape transfer the data is entered into the host system in a pseudo batch mode it is possible with communications lines to develop interactive data transfer to enter the data directly into user accessible database areas.

IAD are currently using public data networks, dedicated lines and satellite links in interactive mode to send and receive CAD data from a variety of client systems. Direct access from remote systems raises the issue of system security and the responsibility for database management in terms of both data revision levels and recording data transfer dates to meet project deadlines. It has proved most effective to isolate a section of the host CAD system for remote access and give local users read only access to the files contained in it; by this method the project management staff can be assured of the integrity of input data records as provided by the supplier.

DEVELOPMENT, INSTALLATION AND UTILISATION OF INTERFACING SOFTWARE

The problem of data standards for transfer of part information is only semi resolved. With a myriad of formats to choose from there remain many permutations of hardware and software for which it is not economic to install interfaces for all but the largest projects. It is to be hoped that the smaller installations will be helped in a realistic manner by the larger client companies for which the required investment is relatively less. Interfacing packages can be expensive to purchase and install; linking this to the fact that CAD software can be revised regularly in such a way that interfaces also need revision, then the whole issue becomes one for constant monitoring, management and investment.

With transfer of data between CAD systems of a different mathematic base it is often necessary to include within the interfacing code some approximation routines. It is essential that the tolerancing levels set by the routines are tested as being sufficiently accurate for the application without being so excessive as to impose too great a CPU demand or to generate output data of excessive volume. In order to simplify interface routines many programmers break down complex entities into more simple forms. Thus if a part is transferred out through an interface and then back to the originating system the resulting part may define the surface correctly but comprise too many simple entities to be manageable. It is not common practice to install regrading routines to upgrade simple surface patches on systems which can support high complexity. Degradation or simplification routines may be essential in certain cases such as transfer between systems with similar mathematical bases but operating to differing maximum orders or degrees of surface. In our experience it is good practice to limit the designers operating the high degree system to the maximum degree of the target system. In this way patch boundaries can be controlled to locations which make sense in engineering terms rather than being established in an arbitrary fashion by a degradation routine.

In Conclusion

The automotive industry has long recognised the potential for CAD applications. It is of particular benefit in the field of surface definition of body panels and has reached quite satisfactory levels of competence. To achieve the aim of total system integration which is desired by the major manufacturers there is still considerable development required. This development is not limited to work on practical aspects such as data transfer formats, but also on the interrelationships between supplier and client.

As these problems have come to light in present applications work they provide additional guides for those involved in research work. It is essential for new process and products to be developed bearing in mind the desire for total system integration and thereby ease the considerable pains of installation. Given the development of populous user groups and initiatives such as ODETTE organised by the SMMT there will be a more positive response for products developed along these lines.

CAD within the automotive industry is an exciting field of work given the constant change and development occuring and I look forward to future developments helping the industry to produce consistently improved products with rewards to both producer and consumer.

REFERENCES

1/ Future of small suppliers - Contractors requirements for electronic exchange of manufacturing data. I. Prod. E. (1986).

2/ J. W. Fellows, Computer Aided Design and Manufacture, Sigma Technical Press (1983).

3/ M. Peltu, Computers - A Managers Guide, NCC (1982).

4/ J. D. Radford, The Engineer and Society Macmillan (1983).

5/ Guidelines for Computer Managers, NCC (1981).

CAE: Fact or Fiction
A Historical Review within an Automotive Company
W.R. Buell
Ford Motor Company, 642 Meadowlane Road, Dearborn, Michigan 48124, U.S.A.

ABSTRACT

The Automotive Industry has progressed a long way in using the computer to solve its complex problems. This is also traced in the rapid growth in the applications of computers to the field of Engineering. CAE, Computer Aided Engineering, is a growing field that started in 1950s and 1960s with the field of structural analysis. After many false starts, the Finite Element Method (FEM), using the Displacement Method, has finally prevailed as the accepted tool for the analysis of many very complex automotive component problems and the problems associated with total vehicle. Upon this FEM methodology has grown the field of CAE and its applications for the engineer today. This paper traces the growth and implementation of the CAE tools within the Ford Motor Company, North American Automotive Operations (NAAO). The paper concludes with some insights into what the future holds for CAE.

CAE MEANING

CAE has traditionally meant Computer Aided Engineering, but if one looks at this definition then what is also included is CAD, Computer Aided Design, CAM, Computer Aided Manufacture and lastly CIM, Computer Integrated Manufacture. CAE in its truest sense is that part of engineering from concept to manufacture that uses the computer in the process to fulfill those requirements. In addition, there has been much confusion about the fact that the "A" in all the CAX, where X can be E, D, or M can also mean assisted, rather than aided. Remember that the "A" should not be interpreted as automated, because this then has the connotation of being like mechanical robots, and that is not what is really meant. There are those that put the major emphasis on the "C" and the rest of the letters are all small. Then there is the variation of "E" as meaning EVERYTHING, and for some this is all encompassing. "E" has also meant evaluation, and that is the meaning that will be used for the remainder of this paper. Thus CAE has the notation of being Computer Aided Evaluation and as such is mainly restricted to the engineering profession.

A currently used definition of CAE is: CAE is the application of computers to evaluate the functional performance of a product or process. CAE involves:

1.) Creating an analytical description or computer "model" of the part, product or process;

2.) Using a computer program to analyze this model;

3.) Interpreting the results to verify the design or identify needed changes to the product or process.

The "birth" of CAE has its roots in the Finite Element Method (FEM) as applied to the field of structural mechanics. In its present form, most of the utilization of CAE can be found in those groups or activities that use the various FEM computer programs (such as NASTRAN and ANSYS). There are many groups within the company that perform that function. On a computer utilization basis, those computer programs represent the majority of computer resources that are generally used by the engineering community. Other computer programs in the field of kinematics, such as ADAMS and IMP, are coming up to a fairly high degree of utilization. But, still since the vehicle is a structure, the utilization of those programs that perform structural analysis, are used the most.

While this definition is appropriate now, it does fulfill the dilemma of the past twenty five years where the CAE was mainly reserved to the Finite Element Method, as originally used for the rapid analysis of structural components. These analyses originally surfaced in the aircraft industry and then spread to the automotive industry. What has also grown into the CAE fold is the entire field of kinematics, that is the analysis of linkages, fluid flow, heat transfer, vibration, and now the field of non-linear analysis in the above disciplines. In the field of dynamics of structures, namely the crash phenomenon, the FEM is currently being used extensively by some engineers in the automotive industry.

CAE THEN

Mr. Peter Marks, of Automation Technology Products, displayed a very interesting chart, table 1, at The Fourth Chautauqua on Productivity in Engineering and Design... The Quest for

Quality, on October 26, 1986 (Reference 1) that he obtained from Creative Strategies International. In this chart are highlighted some major milestones in the CAE evolution. Interactive graphics started in the early 1960s but in the automotive industry we didn't start to use it for development as a design tool until 1967. The chart also shows Timeshare systems beginning in 1965, but again we didn't start using it until the late 60s and then only seriously began to use it in the early 1970s. The point here is that industry hasn't really applied the available tools until five to ten years down the road. We have historically been very conservative in that regard. Another example is the Workstation and its networking. The chart indicates that that began in 1975. In some of the areas of the company they have had workstations for 2-3 years. Again about a ten year lag. The chart continues on into the future, but that will be covered later on in the paper.

Table 2 highlights many of the major CAE capabilities that have occurred over the years at the company. Our involvement with the Finite Element Method began in the mid 1960s when it became apparent that in order to solve many of the complex Gas Turbine component problems at high speeds and high temperatures we needed something more than just the finite difference method and closed form solutions. At first we used the equation solver module of the SAMIS code to solve the finite difference equations that had been generated externally. Two or three thousand simultaneous equations were just too much to solve by hand and besides those routines were the back bone of the FEM code so it was natural to use them. Gaining the confidence in the SAMIS code, we soon started applying it to problems associated with the structure in the automobile. In general we started with small components, like the brake shoes, suspension arms, and ladder frames. Within ten years, as the computers grew in capability, so did the models. By 1976 we were modeling the entire vehicle, a completely trimmed model and running it using the NASTRAN computer program.

In 1972 we made a very strategic decision to drop the development efforts associated with the maintenance and enhancement of the SAMIS code, in favor of using a third party code from The MacNeal Schwendler Company (MSC), called MSC/NASTRAN. We felt that at the time perhaps we should continue both the in house code for experimental and unique applications, and the outside code for the regular production usage of applications. But with the limited available manpower it was decided to completely drop the in house code. In retrospect this is probably one of the most significant decisions that would come to effect the applications of the FEM for other engineers within the company.

Quoting from Reference 2:

Once we had established the technical credibility and capability of NASTRAN, we faced a more significant challenge - how to interface the program with product engineering. Our philosophy up to that time had been that the product design engineer should always do his own analyses since he fully understood the design constraints, loading conditions, and timing requirements related to his components. Frequently, however, the design engineer had not been exposed to FEM prior to a need to use the programs. There are almost 1800 pages of NASTRAN User Documentation; however, it was not written to be self teaching. In such instances, the design engineer preferred to rely on an analyst already familiar with FEM techniques to assure that a good analysis was conducted in a reasonable length of time. Unfortunately, there was no existing company activity which could provide this service.

In 1972 a department was formed with the charter to perform FEM analysis on a request basis. Thus formed the service group that would later become the nucleus for growth to distributed groups when that centralized group could no longer service all the requests that came to them.

The early days, 1965 through 1975 saw the dependence upon the centralized host computer. The FEM analysis was performed on many computers, starting with the Philco 211 and 212 in the mid 1960s and then migrating to the GE 635 computers that later became the Honeywell 6000 series computers. Because of the high computational requirements in the early 1970s, we used outside services including those at Westinghouse in Pittsburg, PA., the CDC 6600s and 7600s, and Cybernet CDC 6600s. When these outside charges started to become significant, it was decided to bring inside that computing, and so, the business computers of the company, the IBM 360 and 370s 165, 155 were called upon to perform the computational capability. This worked well until such time as other priorities within the company compelled the operators running these business computers to delay the execution of the FEM code. Such items as paychecks, and stock holder dividends took the priority, and perhaps rightfully so. It was hard to convince the design engineer that his analysis would be postponed because of the computer running paychecks. It was also during this time that we migrated from the punch cards to the timesharing (TSO on the IBM) operating systems on the host computers.

It was also during this period that we saw a preponderance of graphics devices grow up to support the requirements of the FEM analysts. It began with using the pen plotters connected to the RJE (Remote Job Entry) terminals and in the early days of

CAD development the display of the FEM models on the CRT using the CAD software. The on line plotters soon gave way to the storage tubes receiving their output directly from the host computers. Incidentally, this operation is still quite prevalent today where those with limited access to CAD workstations, still display their results very quickly and efficiently. In parallel to the storage tubes, we also used raster tubes to display the results of using the MOVIE.BYU code from Professor H.N. Christiansen. In early 1972 when we procured the code, we developed remarkable 3D color displays for presentation purposes, and also for quick analysis of the FEM results. Again, this code was third party code, supported and maintained by the third party. However, this particular code required some amount of effort to install on the various host computers. Usually one person is devoted to this vital task of support, including the maintenance and user "help" support. But generally the various structural codes do support their own pre and post processing via graphics, and hence they are maintained by the third party code developers.

During the years 1976-1986 the utilization of the super computers grew. When the first CDC 176 computer arrived at the company in January 1976, the load on it grew very slowly, but in less than a year it became very heavily loaded. Then in January 1985 when the Cray XMP arrived, the heavy users were shifted from the CDC 176 to the Cray, and in less than a year, that machine, also became loaded. By "loaded" is meant that a 2,000 CPU second job will take longer than 24 hours to be returned to the user. During this time, also it must be clarified that the FEM models became more complex. Some of the bigger models exceeded 30,000 nodes and elements. We are still utilizing both super computers, 24 hours per day, seven days per week. Not only did the models become more complex, but more engineers were using the computers. This caused more load on the all ready over taxed computers.

CAE ORGANIZATION

The utilization of CAE within the design organization will never occur unless there is a place within the organization for CAE. It is not enough for one or two engineers to want to do CAE along with their other functions. Regardless of the intention of those one or two engineers when it comes to getting the job done, if they are not well adept at the CAE function, they will probably take an exorbitant amount of time to perform the function;all the while their other functions suffer, at the demise perhaps of the entire design effort. It is management that must want CAE to be utilized within the design process. We have been fortunate within the company to have a very understanding management; one that sees the value of the CAE process and one that feels committed to it. Hence, the organization has reflected this committment.

In 1969, there were but two engineers involved with CAE, specifically FEM; establishing the criteria and procedures required for an effective utilization of CAE. See table 3. In those days, the two engineers were within the centralized engineering Computer Sciences Department; a department with overall computing responsibility for the entire Engineering community. Within that department were the computer operations, advanced computer systems, graphics, and applications activities. For CAE or really just the FEM portion, only two engineers within the entire applications activity were involved. It became evident in these early days, that not only did we have a tremendous capability and potential, but we had one of the best kept secrets: FEM could really solve truly engineering problems, today. In order to train others in the utilization of the FEM code, we started training classes, with the assistance of the training department and a local university. A professor from the university taught the theory portion of the FEM while the two engineers taught the practical side of using the code to solve the real engineering problems.

The effort soon outgrew the applications activity within the computer department and in 1971 a new activity in a sister department was formed. This time it was named the Structural Analysis Section. It is with this group that the tasks began to be identified and specific engineers assigned those tasks. Such items as hardware and software were addressed by specific individuals in addition to those assigned to applications; that is, the practice of using the code to solve real problems. In 1972 this small group of people evolved into a department, and with this the die was cast for the growth of the entire endeavor.

This centralized department served as the focal point for all the major activities involving FEM and the associated accessories, like pre and post processing software, procedures, and the FEM code itself. For the next fifteen years this department served as the major focal point of all the FEM activity. However within six years of the initial formation of the centralized group, other "splinter" groups started to form; chassis, body, electrical, truck (both light and heavy) and scientific laboratory to name but a few of them. Today these decentralized groups comprise the majority of the daily applications being carried out by the company. The centralized group now is attached to a non computer activity and is not serving as the focal point of the FEM activity any more but now acts more like a decentralized activity.

By having the decentralized groups spread throughout the company, more intense, relevant analysis can and is performed on a timely basis. Results can be directly communicated to the requesting engineer by usually walking a few hundred feet to his desk. The proximity to the design engineer is of paramount importance in order to gain the relevance of the design and the modifications as it progresses through the design process.

In a fashion similar to the one centralized group, now these various distributed groups have a teaching and training mission for those they serve. The centralized group performed the teaching and training for the distributed groups. These decentralized groups must now perform similar missions to those design engineers. It is not a simple task. Since the real goal is to encompass more than just the FEM, it now rests on these groups to diversify into more than just FEM. It does throw an added burden on them; one that they must understand and accomplish if these endeavors are to continue and grow.

It is of course the design engineer who must profit from all these training and teaching efforts. There are now some small groups of one or two engineers within a few of the design departments who are becoming the local experts in the CAE technology. This is getting close to the desired results of affecting the design engineer in his decision process. The local experts are responding immediately to changes in the design direction. They appear to be very effective.

It is also very relevant to understand the transformations that have been occurring in the computer hardware industry. It was thought by Mr. Watson, Sr., founder of IBM, that a few centralized computers located in the country, were all that would be necessary to solve industries' problems. Perhaps no more than three or four! Interesting. But the centralized concept was the initial implementation of computers as table 1 depicts. Distributed computing will augment the distributed CAE groups, while the engineering workstation could assist the individual engineer with his specific, daily work tasks relative to CAE.

PITFALLS

There are certainly pitfalls associated with whatever the process. What is desirable to do is to lessen, that is minimize, the effect of the pitfalls. Programs and training courses can overcome them.

The pitfalls of table 4 are perhaps some of the reasons why CAE has not flourished as much as it should. For example, the utilization of the FEM as we know it today can be very complex, and involves many things, such as knowing how to use NASTRAN, knowing how to use a computer (the operating system) and the CAD hardware and software. Each of these items can be difficult and take some time to learn. A lot of these items require extensive skills to be more than of a superficial value. There is also a limited accessibility of the equipment, and expert people that know how to use it.

For the average design engineer, or project engineer, there is infrequent usage of the process, or need to use the process. The design of the component is only a part of the entire process required from concept to customer. As a result, seeing that the engineer doesn't use the process every day, there can be much forgotten, only to be relearned the next time it is needed. Incidentally, in between the times that the process is used, it is possible that the skill required to use some part of the process would change. For example, now the engineer might be required to learn how to use the new super computer, not an impossible task but one that must be undertaken.

It is possible to overcome one of the major hurdles because of the use of third party software. It is the third party's responsibility to port the software to the newer hardware, operating systems, etc. Hence, when one is trained on some software that is portable, the learning curve has been reduced to just that of the new hardware, operation systems, and nothing has hindered the application of the utilization of the software (except perhaps of newly acquired capability because of new functionality of the hardware). These "new bridges" are then maintained by the third party vendor.

Table 5 itemizes the dependences of some hardware and software. We in the CAE community have tried to stay with hardware independence and software that is independent of hardware. The primary step towards fulfilling CAE is the maintenance of these independent strategies. It is important, but there are prices to be paid for such independence. One is perhaps a sacrifice in speed, that is response time, and another is costs. Each CAE activity will have to evaluate the tradeoffs taken for each alternative.

THE FUTURE

Planning for the next five years requires the prediction of CAE market place which implies the level of computer implementation. To predict what the CAE market place will be like is difficult. It could be obtained by extrapolation of current market trends, speculation, or just some good guessing. The future will be very exciting and challenging for all concerned.

In the emerging years, CAE will become much more dependent on CAD, the electronic design. Computers will give us the ability to go from the inception of a design right to the manufacture, without prototype, mockups, and clay models or at least a less dependence on so many prototypes, etc. This will hopefully be going right from "Art to Part."

Many parts are currently being designed on the computer today.

Computer-controlled milling machines can make the parts. If these processes could be combined, months could be pared from most product development programs. But putting the two together has been a major problem. For one thing, the computer aided design system hasn't contained all the data needed to produce an actual part. Also, the way the CAD system describes a part could not be understood by the milling machine. The future holds great promise for solving these problems.

In the next five years, the work place will be electronic and by then, CAD could be the Master. This has many implications, many problems to be encountered, and many solutions to find. The CAE community should be working towards these goals by planning and setting priorities now. CAE software will have to interface quickly with the CAD data base. Standards are being implemented and in a few years there will be better software standards, as shown in table 1.

The CAE industry is working towards the knowledge based expert systems. These systems are an emerging tool of the late 1980's for computer aided decision making. (Another meaning of the term CAD) Such systems capture the decision making processes of experts and allow non-experts to avail themselves of the expertise in the absence of the expert. Decision making in such diverse areas as oil drilling, robot control, medical diagnosis, and shop floor control have been aided by expert systems. These same techniques can be applied to the design and manufacture of the vehicle. It can mean that one could increase the expertness of the individuals which will have implications in the manner and way you will want them to do their day to day business.

CONCLUSION

Is CAE fact or fiction? From the information presented in this paper, CAE is a fact; it does work; it is a productivity tool that in the right hands can produce a superior product. However, as with any technology, if left by itself, it will fall by the wayside. CAE must be cared for, maintained and expanded if it is to continue. CAE - Computer Aided Everything - it is not. It is a lot of things to a lot of people, but not everything to all. CAE tools provide an outstanding evaluation of a product or a process. It is here today. It has been used effectively in the past twenty years. Hopefully it will continue to be used in the years to come.

REFERENCES

1. Marks, Peter, Tutorial #3 "Hidden Issues in Managing the Integrated Engineering Environment", at The Fourth

Chautauqua on Productivity in Engineering & Design.. The Quest for Quality, October 26, 1986.

2. Hamann, W. C., Interfacing Finite Element Methods with Product Design Engineering, ASME 73-DE-24.

Table 1
CAD/CAM/CAE Then and Now

Historical Development of CAE

Hardware	Software
1950 Commerical Mainframe	NC Programming Group Technology
1955 Specialized Line Printers	Wiring Lists Logic Diagrams
1960 Interactive Graphics	USAF Project Sketchpad

1965 Timeshare Systems
    Service Bureaus
    Commerical Interactive
    Integrated Circuit
	Database Technology
	Custom FORTRAN and Algol
	Engineering Programs
	PCB Layout

1970 Storage Display Tube Data Tablet	Mini-based CAE Systems Hierarchical Databases

1975 Turnkey CAD Systems
    Microcomputer
    Raster Displays
    Intelligent Terminals
	IC Mask Design, Simulation & Test
	Relational Databases
	Specialized Commerical Engineering
	Applications Packages
	Workstation Networks

1980 Commercial Array Processors	Scupltured Surfaces Solid Models Kinematic Programs Micro-based CAE Systems Flexible Engineering Systems

1985 Erasable Optical
    Disk Memory
	Engineering Manufacturing
	    Integration
	Expert Systems
	Voice Data Entry

1990 Commercial Flat Panel Displays	Software Standards
2000 Holographic Displays	Flexible Factory

Source: Creative Strategies International
Peter Marks, Automation Technology Products

Table 3
Development of FEM - As An Organization

Year	Event
mid 1960s	Two engineers performing FEM within a computer department.
1969-1970	A staff section within a department is formed.
1972	A department is formed for FEM.
1978	Distributed groups forming in: Chassis, Body, Engine
mid 1980s	Major activities within the company have groups devoted to FEM: Light/Heavy Truck, Scientific Laboratory, Vehicle Development, Advance Vehicle, Chassis, Body, Engine Design Center and other product offices.

Table 4
Pitfalls in CAE

- Lack of training (technology transfer and skill development).
- Lack of sufficient pilot projects to expand breadth of applications.
- Difficulty in using hardware.
- Difficulty in using software.
- Lack of adequate academic courses by employees.
- Insufficient time to perform an analysis.

Table 5
Examples of Hardware/Software Dependencies

| Software located: | Hardware | |
	Dependent	Independent
In house	PDGS (CAD)	TIDE/MOVIE.BYU
Third party	Silicon Graphics	Plot 10 (Tek) NASTRAN, ANSYS

Table 2
Highlights of CAE

1950 . Whirlwind I computer at MIT displays output on a CRT
1962 . Ivan E. Sutherland, Ph.D. dissertation: "Sketchpad - A Man-Machine Graphical Communication System".
1958 . Beginning of proliferation of machine codes for Structural Analysis in U.S.A., especially in aerospace industry.
1964 . NASA initiates NASTRAN code... $3M (completed 1970).
1965 . SAMIS program developed by R.J. Melosh, Ford Philco.
. First CDC/Lundy CAD system for interactive CG.
1967 . Development of graphics as a design tool.
1968 . First courses in FEM - SAMIS.
. SAMIS state-of-the-art with 32 noded solids.
1969 . COSMIC/NASTRAN released.
1970 . Newsletter of FEM prepared.
. NASTRAN code purchased from COSMIC.
1971 . NASTRAN (IBM version) running for general.
. More courses in FEM analysis.
. SAMIS dropped.
. 2D mesh generating code developed.
1972 . NASTRAN level 15 (COSMIC) running.
. CDC/Lundy CAD system displays structures.
. Centralized group formed for analysis.
1973 . Tektronix terminal used to display structure plots.
. IMP developed for 2 and 3 D rigid link mechanisms.
1974 . NASTRAN level 16 developed with non-linear capability.
. Automatic 3D mesh generation developed.
1976 . Interactive 3D modeling available.
. FEM anaysis of '79 Panther program using NASTRAN.
. CDC-175 arrived. Soon replaced by a CDC-176.
1977 . 42 CAD terminal installed.
1978 . Distributed groups formming.
1979 . NASTRAN/MOTRAN preprocessor developed.
1980 . FEM pre and post processor on CAD terminals in running.
1981 . Prime replaces CDC as CAD mini computer for driving CRT.
1983 . Apollo workstations arrive.
1984 . Office Automation equipment available for engineers.
1985 . Cray XMP arrived.
1986 . LAN installed.

Experience with an IEEE 802.3 Local Area Network in a Multi-Vendor Environment
W. Schmatz
MAN Technologie GmbH, Munchen, West Germany

ABSTRACT

The local area network (LAN) techniques may be a strategic approach to some fundamental information processing needs. The MAN Technologie GmbH started the implementation of a local area network for

_ integration of applications
_ access to central processing

The company´s specific situation with data processing equipment of different vendors led to a standard IEEE 802.3 solution with multistar topology on the basis of fibre optic cabling.

From a manager´s viewpoint the following topics are discussed:

_ availability of LAN-products
_ functionality
_ throughput
_ maintainability
_ system price

The result may encourage potential new users.

STARTING POSITION

Company's profile
MAN Technologie is the research and development centre of the MAN group. Here some 1000 engineers and scientists are working on new products, future-oriented projects and state-of-the art technologies.

The company is a leading partner in the European aerospace program ARIANE, with additional developments in satellite engineering. Other products concern wind energy plants with a power output from 20 kW to 1.5 MW. Thanks to the expertise in engine technology a prime contribution to the development of cogeneration plants was achieved. Special knowhow is available to produce gas centrifuges for uranium enrichment.

Information processing applications
A variety of computer assisted tasks run on machines of different vendors . Starting from design, CADAM and CATIA on IBM are used. On a VAX8500 modelling for finite element analysis is undertaken with PATRAN, producing input for NASTRAN and MARC and processing the output for coloured printouts. On the same computer the integrated 2D/CAD-CAM system EQINOX is introduced to generate intermediate drafts of complex mechanical parts and to create interactive the numerical code for NC-machining.

The MAN Technologie operates different minicomputers for 3D-measuring devices and automized testing facilities.

In the commercial field two HP3000 are used, linked to IBM mainframes.

Most applications have longterm histories and were selected to fit the specific needs best. This situation is often reported. The information processing manager's strategic task for integration is well known.

Geographic implication
With respect to the theme, the widespread geographic situation is of importance. The laboratories, production floors and offices are spread over 30 buildings. The data processing is in effect distributed by topological reasons. The economic access to central services has to be organized.

THE LOCAL AREA NETWORK APPROACH

General LAN-functions

There is no general definition for local area networks. The common understanding is: base for open and universal communication even between different systems/equipments. The LAN may serve different objectives:
- linking mainframes
- linking minis and/or personal computers
- supporting terminal access
- supporting the base for distributed processing
- supporting manufacturing tasks
- supporting backbone-functions for linking subnets

The special suitability of an individual local area network supporting one of the above mentioned characteristic or any combination had to be evaluated.

Company´s objectives in LAN application

The primary intention was to utilize the LAN-technique as a technical approach to some fundamental information processing needs. In a first stage, two goals should be reached.

Integration of applications In a distributed environment you need many services to integrate your application. Leading vendors offer today standard software for the following topics:

- Full network service within one vendor´s product line (e.g. within DECNET you get file transfer, remote access, peer to peer communication)
- Fast file transfer between applications on different computers (e.g. transfer of images from a host to a workstation, specialized for animation or fast transfer of IGES-files for CAD-CAD data exchange)

Access to central processing Traditional terminal-connections resemble a tree; the radix is a single computer. Connecting a device via a server to a local area network may change this structure:

- The logical connection to a series of computers within one homogenous system becomes standard (e.g. a printer server may be attatched to a network, to do all printing work for a couple of processors)
- The access of terminals to all computing servers within one homogenous system is standard. Hopefully some terminal server protocols will be enriched by intelligent procedures for access to foreign computers./1/

Using standards
The commitment for using standards is a good practice. When the MAN Technolgie started in 1985 the market leader had not announced his product. Therefore IEEE-standards were investigated.

THREE STANDARD LOCAL AREA NETWORK TYPES

The following table represents the basic characteristics of three standard local area networks. IEEE 802.3 and its predecessor ETHERNET are widely used in thousands of ways around the world. IEEE 802.4, the so called Manufacturing Automation Protocol (MAP), will be operative within the next year. IEEE 802.5 is adopted by IBM as token ring solution. IBM-Products have been available since 1986.

Access methods
Two very different access methods have been established. CSMA/CD uses a stochastic model:
_ a station - ready to send - senses the carrier until the medium is free
_ the station starts sending; there is a risk that another station starts within the signal run-time
_ collisions must be handled.
Token access methods show a deterministic behaviour by enabling and disabling sending permission.

There is a broad discussion about the advantages/ disadvantages of these methods. A comparison is given in /2/.

ADVANTAGES OF A FIBREOPTIC BASED IEEE 802.3 LAN

CSMA/CD-Advantages
Besides the availability, the propagation and magnitude of component-market, one keypoint was decisive. The CSMA/CD-access method permits the coexistence of Ethernet and IEEE 802.3 on the same network. In mid 1986 Hewlett-Packard´s software products followed IEEE 802.3, Digital Equipment´s products followed Ethernet version 2.

Fibreoptic advantages
It was impressive to see, how innovative designers found a way to enlarge the concept of IEEE 802.3. They kept the interface of the medium-attachment-unit exactly as it was. They changed the physical medium completely, enriched the topology and widened the regional application. The following figure shows the conceptual differences.

Standard	Characteristics	Availability
IEEE 802.3	baseband 10 Mbit/sec CSMA/CD bus/multistar	Ethernet version2 1982 IEEE 802.3 1985
IEEE 802.4 (MAP)	broadband 6 MHz token bus (tree)	Version 3.0 1988
IEEE 802.5 (IBM)	baseband 4 Mbits/sec token ring	1986

Standard local area networks

MAN Technologie

252 IEEE 802.3 LOCAL AREA NETWORK

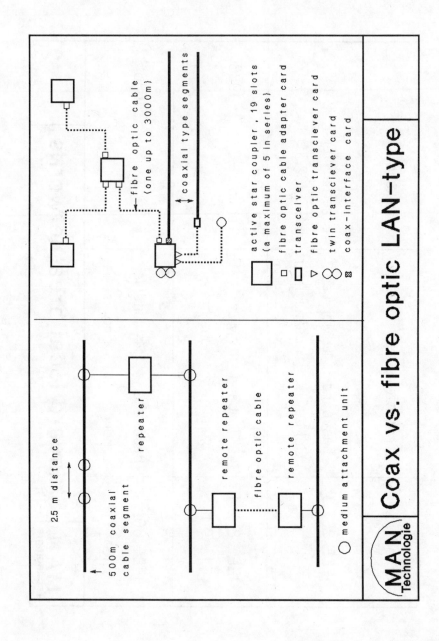

The MAN Technologie decided for the first German supplier, Richard Hirschmann, Radiotechnisches Werk, Esslingen/Neckar, for the following reasons:

No grounding problems The initial network is spanned over two buildings. The coaxial type version demands either for careful grounding or for a fibre optic based remote repeater. Both solutions are expensive. The latter one limits the application, if multiply applicated.

The fibre optic solution is by its nature free of this problem.

Topological flexibility The conventional way, to link all equipment, is suitable for small installations. A better approach is the topology, which IBM has elected for its cabling system. The multistar concept of Hirschmann supports this idea. In every building one star can serve all necessary connections. Additionally transitions to the coaxial type are possible at any star coupler.

Wide range Each segment of the coaxial based type is limited to 500m. The maximum of 2 repeaters limits the range. On the optical medium the deformation of the digital signal is smaller. The range extends to 4500m including one single line of 3000 m and a maximum of five stars in series.

Repeater free base network The absence of repeaters cuts the signal run time. Insofar it reduces collisions by a small amount.

EXPERIENCE AND PERFORMANCE

To represent the latest experience, dates for special cases and mixed application will be given on presentation.

INSTALLATION AND MAINTAINABILITY

Fibre optic installation
All installation work was done with company´s own staff. A skilled person will learn the mounting of connectors within a day. There is only one limiting factor: for repair of connectors about two hours are needed, mostly for hardening of adhesive.
   The transmission should be measured. For small installations a visual inspection is sufficient.

System installation
In all cases the system services were first proven with the vendors coaxial type. The switching to the fibre optic type was a task of a few minutes. In one case the light emission had to be adjusted; this is possible in three stages by jumpers.

Maintainability
All cards have LEDs for the following indications: "power", "data accepted" and "collision detection". They are helpfull for installation and operator control. For performance measurements and problem solving special LAN-analysers are necessary. Personal computer based analysers are the cheapest solutions.

SYSTEM PRICE

On the pricelist of the manufacturer mentioned above the following simple configurations are calculated (rounded prices in DM, base 1986):

Point to point configuration, 300 m distance
_ 2 transceiver	5.500 DM
_ 300 m fibreoptic cable	3.000 DM
total	8.500 DM

Start to star, 4 direct, 4 remote interfaces, 500 m cable
_ 2 active star coupler	10.000 DM
_ 2 twin transceiver cards fitting into the star coupler	3.000 DM
_ 6 fibreoptic cable adapter cards	10.000 DM
_ 4 fibreoptic transceiver	11.000 DM
_ 500 m fiberoptic cable	5.000 DM
total	39.000 DM

The prices compare well to a network with expensive remote repeaters.

CONCLUSIONS

From a manager´s viewpoint the MAN Technologie has introduced a proven local area network with an advanced technique. The goals could be reached. The solution is economic.

The choice for this individual solution is not a final one. It is obvious, that new developments respect existing standards. The greater the installed basis, the greater the pressure, to build gateways to other solutions from the beginning.

Hopefully this paper may encourage potential users.

REFERENCES

/1/ Raabe, Georg-Peter, (1986) Planet in der Universität Passau. State of the Art, 2/86 63-74

/2/ Suppan-Borowka, Simon, (1986) MAP Datenkommunikation in der automatisierten Fertigung. Datacom Buchverlag, Pulheim

/3/ Holler, E. (1986) Ein LAN für ein Großforschungszentrum: von der Planung bis zur Realisierung. Datacom, Pulheim, 2/86, 4-21

GLOSSARY

CADAM	System for computer aided design, Release 19.1, developed by Lockheed
CATIA	System for computer aided design, Release 20.2. developed by Dassault
CSMA/CD	Carrier Sense Multiple Access with Collision Detection; access method for IEEE 802.3 and Ethernet
Ethernet	Local area network, developed by Digital Equipment Corp., Intel and Xerox Corp.
EQINOX	EQINOX 7000: System for computer aided drafting (DRAFT) and interactive generation of NC-code (NCG), Applicon Schlumberger GmbH, Frankfurt
IEEE	Institute of Electrical and Electronic Engineers, New York
MAP	Manufacturing Automation Protocol: Local area network, introduced by General Motors Corp.
MARC	MARC: General purpose finite element program, Revision K2-2; Marc Analysis Research Corp., Pala Alto, California
NASTRAN	MSC-NASTRAN: Finite element program, Version 65; MacNeal-Schwendler-Corp., Los Angeles, California
PATRAN	PATRAN II: Finite element pre- and postprocessor, Version 2.0A; PDA Engineering, Santa Ana, California

Applications of Voice Recognition Techniques in the Automotive Industry. (Austin Rover Group, England)
R.P. Anderson
Austin Rover, Birmingham Operations, England
K.D. Gill
Austin Rover/Warwick University Teaching Company, England

1. INTRODUCTION

Speech recognition and voice response technologies have now advanced to a point where practical applications can be made in the manufacturing environment. This was the opinion held by Austin Rover after extensive investigation of technical literature and also feasibility studies carried out in early '86. During this survey, most of the products and technology reviewed were only suitable for laboratory or office environments. This proved to be a setback as very few companies had experience of implementing voice systems within the shopfloor environment.

In several cases, vendor's solutions were not practical due to hard wire links between the computer and user. This would reduce the necessary mobility of operators, therefore eliminating one of the benefits provided by voice technology. Another problem was the use of discrete word recognisers which provide the user with a totally unnatural dialogue. One company, Voice Systems International Ltd based in Cambridge, England, provided a solution which overcame both these problems but at that time was still not the required 'turnkey' solution. On offer at that time was VSI's DataVox Qualitator product which used a Votan VPC 2040 continuous recognition and voice output board housed in an IBM.XT. The voice interface for radio was achieved by means of a VSI custom built interface card to make full use of low powered simplex radio equipment.

Austin Rover's requirements then led to the development of the system software to enable the collection of fault reporting data.

2. INITIAL TRIALS

Early investigations into voice applications and capabilities indicated that it was ideally suited to the Quality Audit areas within ARG's manufacturing plants. Austin Rover's confidence in the technology was then re-inforced by the purchase of a single DataVox Qualitator unit which was evaluated on the Mini Model assembly tracks. These early trials taught the Project Engineers that each inspector could use the system with varying degrees of success.

Two inspectors were chosen at random, and taken to an area remote from the shop floor. The principle of the equipment was explained to them s that they could appreciate the technology they wer using. This proved to be invaluable as a genuine interest soon developed and they were not just dealing with "another machine".

Before the inspector uses the system, he must first of all train it to recognise his voice. Thi involves the inspector training the system in his method of speaking each word. Each word is repeated and stored as a template, which is then compared with the spoken word when the system is being used.

The first training session lasted for around two hours, during which the natural inhibitions of speaking into a microphone were overcome.

During this session the inspector tended to try to over - pronounce words in order to "help" the computer.

The initial training passes for both operators were generally poorer than expected. Discomfort, coughs and nerves caused a number of words to be trained poorly. In order to determine whether a particular user's vocabulary was trained well, each operator performed several verification passes. Words which performed badly were simply re-trained. These tended to be the single syllable words, although one of the inspectors did have a pronounciation problem on two of the multi-syllable words. This was solved by simply changing the word, one of the many benefits of voice recognition techniques.

After several runs through the inspection sequence for the Rover 200 series models, the inspectors were retrained using the radio and headset. This introduces further variables in the form of headset mounted microphone and a voice operated switch. The latter activates the radio during the first few micro-seconds of the spoken word. The use of a radio alleviates the need for trailing wires which would inhibit the user's mobility, and prevent use within the cabin of the vehicle. (Figure-1). Training was again repeated, taking around fifteen minutes to train 36 words twice. The vocabulary used consists of the model type, the fault descriptions, numbers and operating commands.

The two inspectors were then taken to the Quality Audit Department to enable them to become familiar with the vocabulary and the structure of the interaction. The users began to revert to their normal mode of speech, so that recognition became inaccurate. At this stage the inspector's templates were again re-trained.

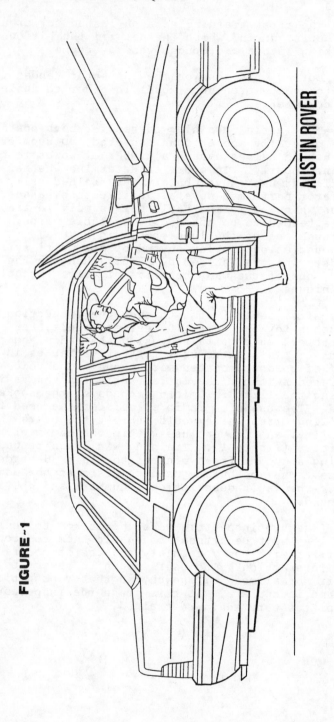

FIGURE -1

This was due to the inspector speaking 'mechanically' during the first training cycle.

On completion of this second training session the system was installed on the Mini Model Trim Tracks. When using the system on the shop floor the inspector spoke louder due to the increased level of background noise. The templates trained were not representative of the way the operator spoke on the shop floor and we therefore retrained the operators again to reflect the shop floor environment.

3. LINE TRIAL EXPERIENCE

The system was used by both inspectors, on a daily basis for a period of three weeks. During this time, it was confirmed that the technology could be used within the assembly track environment, although several changes and modifications would have to be taken into account.

Both inspectors achieved fairly good levels of recognition accuracy (between 90 and 95 %), the main problems being:-

- during training the volume at which he trained was significantly different to the way he spoke on the track.

- when problems were encountered the inspector began to over - pronounce to try to achieve the correct recognition.

- even though he was fairly relaxed on the track he was extremely tense whilst training.

- there is a definite rhythm which is required to achieve the high accuracy required.

- problems were encountered with certain key words of the vocabulary.

The acceptance of new technologies (especially in the manufacturing environment) is not a problem which is unique to speech recognition. The user develops his or her own opinion based upon their interface and contact to the system.

For this reason the following points were noted from the trial:

- Speech systems should be fairly adaptable.

- The user is disciplined in performing the designated inspection checks.

- The user should feel confident in the transaction. Reliability is important.

- Feedback must be simple and almost instantaneous.

- On-line guidance must be simple and consistent.

- Dialogue must be structured to avoid confusion.

Using this experience a system was designed for use on the final inspection line of the new ROVER 800 series executive car.

4. ROVER 800 INSPECTION SEQUENCE

An inspection shift is commenced by the inspector identifying himself to the system by a two stage log-on procedure. The first entry is the inspector's identification code followed by his password. The system is able to withhold 64 'voice prints' for the identification stage, and compares the input to this library. If, the result is incorrect, the inspector is able to try again. Once the user is identified, the program and operating vocabulary is loaded from the hard disc.

On arrival of the first car to the inspection zone, the build card is located by the inspector. Information which uniquely identifies the model derivative etc... is then entered by a barcode reader which is interfaced to the computer. If the information entered by the barcode reader is incorrect then an entry can be made via the keyboard. From this description a generic checklist, which is relevant to the vehicle to be checked is prepared.

The inspection proceeds by a basic dialogue of prompt, response and verification between the operator and computer. The computer transmits a check to the operator that is similar to a low quality tape recording. During the initial investigations and trials, it was decided that this was more acceptable than the currently available synthesized techniques.

The inspector can respond to each prompt in one of three ways. The item can be passed, failed or the prompt can be repeated. If the item is passed, the prompt for the next check is heard. On failing the item the inspector can choose from fifteen descriptions of faults. After the faults are recorded the inspector terminates the input. The faults are then repeated for verification and if correct the inspector moves on to the next check.

The computer instructs the inspector, on which valid faults are applicable to any particular check, upon request. Likewise, if an invalid fault is entered, the system will correct the inspector. At the end of the sequence, all faults which have been logged are passed to the next system where they are stored. These are then added to when the car enters the second zone. At the end of the inspector's cycle, a fault file is produced which is printed and attached to the vehicle build card, any faults found at this stage being rectified.

At the end of each shift statistical information is available so that trends in the manufacturing process can be evaluated. This can also be requested 'in shift'.
(Figure-2, illustrates a typical voice driven inspection station as currently used for quality checks on the Rover 800 series models.)

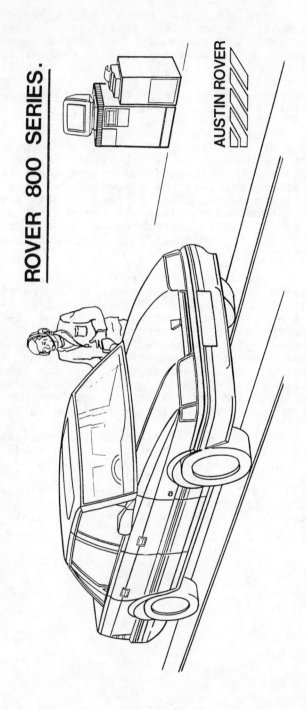

FIGURE - 2

5. PERFORMANCE TO DATE

The system now is very different to the pilot installation as used on the Mini Model at the start of the evaluation process. Experience gained by both Austin Rover and VSI during the trial period, enabled fine tuning of the human interface. One area where a great improvement has been made is training of the end user.

This is the result of the standardisation of the vocabulary used and also the training methods. It now only takes about 30 minutes initial training before the user can operate in a fairly confident manner. Subsequent re-training only takes about 10 minutes and the number of re-training sessions required varies between users. So far, twelve inspectors have been trained, all of whom have achieved a 95% accuracy of recognition.

The success of voice technology within Austin Rover is due to a variety of reasons. One of the most important is the high level of acceptance by the inspectors (end-users) and their shop floor collegues. This is due mainly to a gradual introduction, which was seen as a development phase. During this time, the system was designed and structured to meet the requirements of the end user. During all the line trials, inspectors who have no knowledge of computers have been trained to not only speak to the computer, but also load their voice templates, prepare quality data, and most important of all, provide constructive criticism.

Selection of inspectors is also important during development as it is very easy to 'kill' a project at this early stage. Once experience is gained any user can be successfully trained.

The next phase of the ROVER 800 project involves installing further units in all Quality Assurance areas.

6. BENEFITS OF SPEECH AS A DATA CAPTURE METHOD

There are a number of areas where the use of speech as the interface to a quality system gives appreciable benefits.

The most important of these is the real time nature of speech recognition systems. As compared with conventional methods whereby data is analysed several hours or even days after the event, then the almost instantaneous interrogation of the system offers a real time picture of the process. This has the result of reducing the time to identify, isolate and rectify faults, therefore reducing costly rework and rectification.

This has the effect of contributing to improved quality as fault information can be relayed to the zone which created the problem. This information can also be relayed to product designs, production control, and quality so that the problem can be isolated and immediate effective action can be taken.

It also makes the user's job easier and can be faster as the system tells him what to do but does not dictate the pace to him. In some cases, several written entries had to be made. These have now been eliminated, as the inspection process is now controlled by a computer. This latter fact will now ensure that no cars can be released until the computer output is passed off.

The use of the headset and radio allows the inspector to use his hands and eyes more frequently, making him more effective and efficient, (Figure-1).

One benefit which was realised after the trial was completed, was the standardisation of the description of the faults. This allows quality data to be analysed over a larger range to identify fault trends. (e.g derivative range of vehicles).

7. FUTURE DIRECTIONS

Several proven applications now exist in the United States of America which include materials control, materials handling (warehousing) and n/c data preparation.

The experience gained, by Austin Rover over the past year on voice technology, has led to the further introduction of a second voice recognition facility for the Metro/Mini models (Figure-3), on paint inspection within all Quality Assurance areas. This application is totally different to the system employed for the ROVER 800 models, due to the nature of the inspection sequence. It has therefore been noted that there is no one solution which covers all applications and so sufficient time must be spent at an early stage to ensure project success (evolution not revolution).

Other applications are being investigated (e.g machine tool monitoring and pallet loading).

Austin Rover's experience to date proves that voice recognition is a realistic alternative means of data capture. For the ROVER 800 models, it is considered that the factory floor inspection system will:-

- ensure that all inspection checks are completed
- ensure that any faults are found
- increase product quality levels to higher standards.

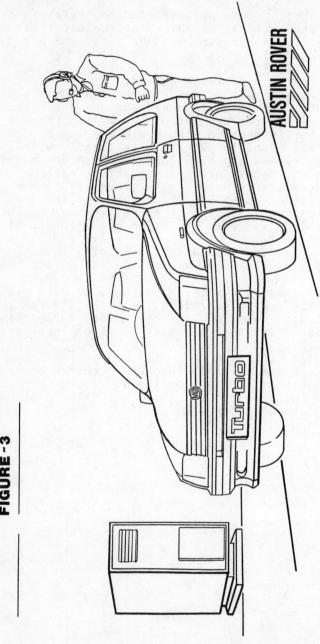

FIGURE -3